T0332356

SINGULARITIES: LANDMARKS ON THE PATHWAYS OF LIFE

This brief but deep book presents a sophisticated consideration of the key steps or bottlenecks that constrain the path to the origin and evolution of life. Christian de Duve, a pioneer of modern cell biology and Nobel laureate, gives in this book a contemporary response to Erwin Schrödinger's tremendously influential *What Is Life?*, which 60 years ago influenced many of the pioneers of molecular biology. Christian de Duve offers shrewd insights on the conditions that may have first called forth life and surveys the entire history of life, using as landmarks the many remarkable singularities along the way, such as the single ancestry of all living beings, the universal genetic code, and the monophyletic origin of eukaryotes. The book offers a brief guided tour of biochemistry and phylogeny, from the basic molecular building blocks to the origin of humans. Each successive singularity is introduced in a sequence paralleling the hypothetical development of features and conditions on the primitive earth, explaining how and why each transition to greater complexity occurred.

Christian de Duve is Andrew W. Mellon Professor emeritus at The Rockefeller University, Professor emeritus at the Catholic University of Louvain and Founder-Administrator of the Christian de Duve Institute of Cellular Pathology. In 1974 he was jointly awarded the Nobel Prize in Physiology or Medicine, with Albert Claude and George Palade, and still conducts groundbreaking research into the origins of life.

Singularities

LANDMARKS ON THE PATHWAYS OF LIFE

CHRISTIAN DE DUVE

CAMBRIDGE
UNIVERSITY PRESS

Shaftesbury Road, Cambridge CB2 8EA, United Kingdom

One Liberty Plaza, 20th Floor, New York, NY 10006, USA

477 Williamstown Road, Port Melbourne, VIC 3207, Australia

314–321, 3rd Floor, Plot 3, Splendor Forum, Jasola District Centre, New Delhi – 110025, India

103 Penang Road, #05–06/07, Visioncrest Commercial, Singapore 238467

Cambridge University Press is part of Cambridge University Press & Assessment,
a department of the University of Cambridge.

We share the University's mission to contribute to society through the pursuit of
education, learning and research at the highest international levels of excellence.

www.cambridge.org
Information on this title: www.cambridge.org/9780521841955

© Cambridge University Press & Assessment 2005

First published 2005

A catalogue record for this publication is available from the British Library

Library of Congress Cataloging-in-Publication data
de Duve, Christian.
Singularities : landmarks on the pathways of life / Christian de Duve.
    p.   cm.
Includes bibliographical references (p.).
ISBN 0-521-84195-X
1. Life–Origin.   2. Evolution (Biology)   3. Singularities (Mathematics)   I. Title.
QH325.D418   2005
576.8´3 – dc22   2004054761

ISBN   978-0-521-84195-5   Hardback

Contents

Foreword

This book was not meant to happen. When I finished *Life Evolving* (de Duve, 2002), I resolved not to write another book. In that work, and in the three that preceded it (de Duve, 1984, 1991, 1995), I had written all I had to say on the subject of life, including its nature, origin, and evolutionary history, up to and beyond the advent of humankind, and even its cosmic significance. In fact, I made the embarrassing discovery, when called upon to reread some passages of these books, that I tended to repeat myself, sometimes using almost the same words – an ominous sign indeed.

I changed my mind and began writing this book when I discovered that, in trying to reach a wide readership, I had buried a number of scientific points that I felt to be significant and original within more general expositions designed for lay readers. The message I wished to convey had been blurred, even misinterpreted as reflecting the pursuit of an ideological agenda (Szathmary, 2002; Lazcano, 2003).

This realization has prompted me to clarify my thoughts, remove extraneous material, and formulate the result concisely for a scientifically literate readership. My aim is to reach not just my fellow biologists but all those scientists, from physicists, cosmologists, and geologists to naturalists and anthropologists, who share an interest in the place occupied by life, including our own, in the cosmos. For this purpose, I have, with apologies to the experts, summarized once again the key features of life, much in the line of my earlier *Blueprint for a Cell* (de Duve, 1991), but this time with a focus on *singularities*, by which I mean events or properties that have the quality of singleness, uniqueness.

The history of life is marked by a large number of such singularities. All known living organisms, be they microbes, plants, fungi, or animals, including humans, are descendants from a single form of life. All are constructed with the same building blocks and combine these into their characteristic constituents by the same biosynthetic processes. All carry out the same metabolic reactions and rely on the same mechanisms to derive energy from their environment and convert it into useful work. There are differences, of course, depending on the nature of the substances utilized, on the source of energy, and on the kind of work performed. But the basic processes are the same. All known living organisms use the same genetic language; they obey the same base-pairing rules and, with rare deviations of recent occurrence, conform to the same genetic code. Behind the enormous diversity of the biosphere, there clearly lies a single, fundamental blueprint.

Considering later stages of evolution, we find that all eukaryotes are derived from a single ancestral cell. Similarly, land plants, fungi, and animals are each monophyletic, that is, offshoots of a single founding organism; so are the members of each class or family, as abundantly illustrated by cladistics and confirmed by molecular phylogenies.

Often taken more or less for granted, these singularities require an explanation. Looking for such an explanation may reveal some valuable facts concerning the nature of life, its origin, and its evolution. It may also help guide our explorations in search of signs of life elsewhere in our galaxy and beyond.

Such is the purpose of this essay. References to the literature are sparse and largely restricted to recent publications. By necessity, the book is rooted in biochemistry, molecular biology, and cell biology, but in a manner designed to make these disciplines accessible to any reader with some familiarity with the language of modern science and, perhaps, to encourage some to take a more active interest in the disciplines and to acquire the basic notions without which a true appreciation of the significance of life in the universe is not possible.

The book covers a wide area, exceeding by far the domain of my own competence. Several friends and colleagues have helped me make up to some extent for these deficiencies. In particular, Nicolas Glansdorff, Patricia Johnson, Antonio Lazcano, Harold Morowitz, Miklos Müller, Guy Ourisson, Andrew Rogers, and Günter Wächtershäuser have read all or part of the original manuscript and contributed valuable information, comments, criticisms, and suggestions. I am deeply grateful to them but must claim sole responsibility for the final text, including its mistakes and idiosyncrasies, especially in that I have not always followed the advice with which I was favored. I am also indebted to Jeffrey Bada, Johannes Hackstein, Arthur Kornberg, and William Martin for providing me with some of their recent publications. I owe a special debt of gratitude for the invaluable assistance of my friend Neil Patterson, who has, once again, waded through my prose and helped me trim it of unnecessary words and clarify its contents. I also acknowledge with thanks the efficient and constructive collaboration of Peter Gordon, Alan Gold, and their colleagues at Cambridge University Press and of Doug English at TechBooks. Also, I acknowledge with emotion the memory of my devoted and talented Karrie, who helped me in all my writings for more than thirty years.

On Christian de Duve: An Editor's Appreciation

Art and science are all of a piece in the life and work of Christian de Duve. I have had the privilege of serving as his editor on five books in 25 years. I have dined with him at many tables, worked with him at many desks, and witnessed the passion and grace of his labor to understand how life works and how it came to be. In all this time and through all these tasks I have watched the force of his intellect convert to words on paper his deep knowledge and intuitive sense of the most basic biological processes, giving the reader sweeping views of life evolving from molecules to the millions of species with which we share this planet.

It all looks easy when the book is bound – his narrative unfolds in a compelling way, the argument grows in strength, a convincing theme emerges, and its variations weave across the pages with thrilling strength and subtlety – but I've been close enough to see how he does it, and it's hard work, strenuous and painstaking, the kind of slow and heavy effort that we sensibly call a labor of love. One would not do this work just for money; it's way too tough; or even for fame, often a more seductive lure. So it *is* love, I think, that drives de Duve to write these books, just as it was love (the love of truth, however corny that may sound) that kept him at the laboratory bench for more than five decades of exacting science. That's part of what I mean about "art and science"; both, when done at the highest level, produce a piercing effect on the human mind, an illumination.

I began my editorial work with de Duve in 1980. (By the way, my so-called *editorial work* must be seen as thin soup when placed against the power of his contributions; this is not a collaboration between equals, far from it.) He had decided to write a book based on his notes for the 1976 "Christmas Lectures on Science" at Rockefeller University, a book to be called *Guided Tour of the Living Cell*. At that time, my colleagues and I at W. H. Freeman – under the bright and enthusiastic guidance of our boss and mentor, Gerard Piel, President of the parent company, Scientific American – had just begun a new imprint, *Scientific American Books*, and a new series, the *Scientific American Library*. We launched the series with *Powers of Ten* (about the relative size of things in the universe) by Philip and Phyllis Morrison and Charles and Ray Eames. This book gives a wonderfully vivid look at the effect of adding a zero. In an exponential series of images and words, the Morrisons and the Eameses take you, one order of magnitude at a time, each on its own two-page spread, from a picnic in Chicago at 10^1 meters up to 10^{25} meters (one billion light years), where space "looks" empty, and down to 10^{-16} meters, the realm of quarks.

Powers of Ten was a big hit, and it set the intellectual mood and established standards of content and design that we would have to sustain in order to meet the goals set by Piel and his Board of Directors. Enter de Duve. As I say, he was already busy making his lecture notes into a book, not just a scholarly monograph of the sort that scientists of his stature produce routinely for their peers, but a more ambitious enterprise, a survey of the entirety of cell biology that was meant to catch the interest, and inform and entertain, ordinary folks in the open world, the *trade*, as they say in the publishing business. This was the sort of book we at Scientific American needed in order to match the marketing impact gained by *Powers of Ten*. But was this making sense? Who is de Duve? He's in a white coat at Rockefeller University, a Nobel Laureate, a high-powered scientist whose native tongue is French, whose writings up to this point have been academic treatises on the biochemistry and bioenergetics of cellular transactions that only his sophisticated cohorts would trouble to read.

My colleagues and I met de Duve in his offices at Rockefeller University. We told him about our new series, walked away with a chunk of his manuscript, read it, saw that he had what it takes, signed him up, and then the action took off. This was a tiger, and he had *us* by the tail.

First-rate authors in the fields of science are particularly demanding, and for good reason. Science is not for the faint-hearted; scientists demand a lot from themselves and each other – acuity, rigor, integrity, and horrendously hard work – so they expect a publisher to do its tasks in like manner. Such authors, of course, are almost always disappointed, so it is no surprise that de Duve was promptly exasperated. We had assigned one of our most experienced editors, Patty Mittelstadt, to shepherd de Duve's project through the developmental process. Suffice it to say that, despite her muscular will and years of experience in the editorial trenches, de Duve managed to have her in tears on several occasions. A disputed semicolon could cause a storm.

Although our colleague did a strong, professional job under taxing circumstances, de Duve served, essentially, as his own developmental editor; and he worked with the artist, Neil Hardy, in so diligent and scrupulous a fashion that Hardy could render de Duve's "rough" sketches – remarkably clear delineations of molecular and cellular forms and interactions, all to scale, with exact bond lengths and angles, every image color-coded – without having to worry about accuracy. This "rough" art was correct.

It happens now and then that an author becomes a kind of gift to the publishing house, a performer of such astonishing ability that the publishing team – managers, editors, marketers – end up playing, are only allowed to play, a perfunctory role in the process. In a half-century of publishing, I've received just three such "gifts," at three separate houses: at W. A. Benjamin, Inc. (now Benjamin-Cummings) it was James D. Watson's *Molecular Biology of the Gene*; at Worth Publishers it was Albert Lehninger's *Biochemistry*; and at Scientific American Books it was de Duve's *Guided Tour of the Living Cell*. In each of these cases, the author gave so much skill and attention to details of content, level of discussion, writing style, illustrations, and all the other elements that shape such a book and determine its success that the publishing team had only to handle mundane procedures of copyediting, design, page layout, printing, binding, warehousing, marketing, shipping, billing, and collecting. These three books jumped off the shelves. They were bought, not sold. (In my experience, only the rigorous and brilliant Michael Munowitz has achieved this level of authorial perfection, and his gifts went to houses where I was not employed.)

"*Tour*" sold 70,000 copies in a couple of years, a best seller in the trade market for science books at this level, and it demonstrated de Duve's membership in that small and increasingly precious pantheon of contemporary scientists who have the rare ability to write for the general public. This is a serious talent, a critical contribution to cultures around the globe. We live in polities captured by *belief*. Belief is a golden virtue; merely by asserting beliefs one can win huge support in the American political system. But believing is easy, and knowing is hard, and it's knowing that matters most. Those scientists (Peter Atkins, Andrew Berry, Richard Dawkins, Christian de Duve, Jared Diamond, Michael Munowitz, Steven Pinker, Matt Ridley, Oliver Sacks, James D. Watson, Steven Weinberg, Edward O. Wilson, et al.) who can write with clarity and balance and evocative power about that which is so, about the nature of physical and biological reality, from matter and energy at the Planck length to the edge of the visible universe and beyond, who can show what we know and don't know and how knowing itself is achieved and how it must often be seen as tentative and open to question, those scientist-artists, for that's what they are, become an essential human resource, national treasures whose works help wean us from the comfortable obscurantism of convictions born of belief.

I didn't do much work on this first book by de Duve (my main contribution being the engagement of Malcolm Grear, the great American graphic designer, who invented the visual character of this two-volume work, a stunning physical object in its own right), but over the next many years Christian and I together devised what, for me, became a delightful set of editorial rituals. Three great de Duvian works emerged during this period: *Blueprint for a Cell* (Neil Patterson Publishers, 1988, an ephemeral imprint later purchased by Viacom and metabolized within the body of its huge and flourishing subsidiary, Prentice-Hall), *Vital Dust: Life as a Cosmic Imperative* (Basic Books, 1995), and *Life Evolving: Molecules, Mind, and Meaning* (Oxford University Press, 2003). There were lunches and dinners and wine and good cheer, during which outlines were discussed, schedules set, and contract terms constructed. There were our side-by-side examinations of my modest editorial adjustments, mostly removing Gallicisms, for he writes in French and English with equal facility, and sometimes an idiom from one would invade the other. And then there were the Nethen Episodes, week-long sessions of intensive, sharply focused concentration on a "final" manuscript, a wrap-up to ready the work for printing.

Nethen, just beyond the edge of Brussels, is de Duve's hometown, and it is here, in bucolic serenity, that the value of ritual is fully in evidence.

During these sessions, the daily regimen begins at 6 AM with about twenty minutes of swift swimming in the backyard pool. This is not fun; it's humorless aerobic effort, length upon length. Next comes breakfast served by Janine de Duve, his stylish, beautiful, and loving wife of 60 years, an accomplished artist and a woman of high intelligence and the native honesty to express her views with sharp, good-humored precision. Then it's upstairs to our respective stations; de Duve to his study, I to my room. We work separately until noon. Then lunch (and sometimes a walk in the woods), then work, again separately, until the late afternoon, when we meet to consider together what he's done and I've done. At last, cocktails at 6, dinner (always entirely satisfying, often amazing), then, oddly, an interlude of television – the news and, usually, a favorite quiz program broadcast from Paris – and so to bed.

This kind of clockwork action is highly productive and may well be a common procedure among research scientists. For de Duve, even now, in his 86th year, it seems essential; he has a driving need to be productive, to keep at it. But all is not severity and discipline. Every interlude between periods of hard work is a time of great pleasure for him. When he's in New York City (throughout most of his long career he has spent half of each year at Rockefeller University, NYC, and the other half at the de Duve Institute of Cellular Pathology, Brussels, Belgium), he walks briskly, weather be damned, from his apartment on Central Park West, across the park, and over to the shore of the East River, York Avenue and 66th street, and bounds up four flights of stairs to his office. This walk is savored: a time to think, a time to enjoy the physical patterns of life itself. Lunches and dinners are likewise times to satisfy his appetites for food and wine, laughter and conversation, and the enduring delight he finds in the company of friends and colleagues. He makes an excellent friend, and he is a friend to many.

This book, *Singularities*, he tells me is his swan song. But this can't be so. He told me that same thing two books ago, when he was beginning *Vital Dust*. How can we afford to let de Duve retire? In any case, such a condition in no way suits his nature, for which we must all be truly grateful.

Neil Patterson

The Mechanisms of Singularity

In theory, a number of different explanations could account for a singularity. Schematically, I distinguish seven different, not necessarily mutually exclusive kinds. Represented schematically in Figure I.1 (printed on the front endpaper of this book), these different kinds of singularities are, in the order of decreasing probability:

Mechanism 1. Deterministic Necessity. According to this interpretation, things could not have been otherwise, given the physical–chemical conditions that existed. Most physical and chemical phenomena belong to this category. They obey the laws of nature in a strictly reproducible fashion. Only at the subatomic level does quantum mechanics allow for some uncertainty. Life is not affected by events at that level, except, according to a theory proposed by some investigators but far from unanimously accepted, in the brain–mind connection.

Mechanism 2. Selective Bottleneck. This mechanism applies to any situation where different options are subject to an externally imposed selection process that allows only a single one to subsist. The most familiar such situation occurs in Darwinian natural selection, where different organisms compete for available resources within ecosystems and the organism most apt to survive and reproduce under prevailing environmental

Table I.1. *Converting chance to necessity.* A few examples illustrating the number n of opportunities required for an event of probability P to have a 99.9% chance of taking place. Calculations (see text) are based on the following formula (in which P_n is the probability of the event's occurring if n trials are made):

$$P_n = 1 - (1 - P)^n$$

Game	Probability P_n for $n = 1$	Value of n for $P_n = 99.9\%$
Toss of a coin	1/2	10
Toss of a die	1/6	38
Roulette (one zero)	1/37	252
Lottery (seven digits)	$1/10^7$	69×10^6
Point mutations	$1/(3 \times 10^9)$	20×10^9
(replication errors)	per cell division	divisions

conditions ends up predominating. Many other instances of a basically similar process are encountered or can be created, depending on the nature of the competing entities and on that of the selection criteria. Thus, as is done in many bioengineering experiments, replicating molecules subject to mutations may be forced through a selective bottleneck rigged so as to let through only those molecules that exhibit a predetermined catalytic activity, eventually singling out the most efficient among the retained catalysts.

By definition, a selection process of this sort is restricted to the actual variants that are subjected to it. An organism or molecule fitter than the one selected may be possible. But if the required variant is not supplied in the first place, it obviously cannot be selected. When, as is often the case, variants arise by chance, the likelihood that a given variant will be offered for selection depends on the number of opportunities this variant is given to occur, relative to its probability.

This relationship is readily calculated (de Duve, 1995). Let P be the probability of the event; then the probability of its *not* occurring is $1 - P$ in a single trial and $(1 - P)^n$ in n trials. Thus, the probability P_n that the event will actually occur if it is given n opportunities to occur equals $1 - (1 - P)^n$. A few representative values calculated by this equation are shown in Table I.1.

What this calculation is meant to illustrate is that *chance does not exclude inevitability*. Even highly improbable events may be made to happen with near-certainty if enough opportunities are provided for them to happen. As a rough rule of thumb, multiply the reciprocal of the probability of the event by a little less than seven, and you have the number of opportunities needed to give the event a 99.9% likelihood of taking place. As indicated by Table I.1, even a seven-digit lottery number will come out in 999 cases out of 1000 if 69 million drawings are held. This fact is of little interest to buyers of lottery tickets but has deeply meaningful implications for the history of life. It allows for the possibility of *selective optimization* under given circumstances, if the situation is such that the range of possible variants can be explored exhaustively. We shall encounter a number of instances where this may indeed have happened.

Mechanism 3. Restrictive Bottleneck. This term refers to a situation in which internal constraints, imposed, for example, by the structures of genomes or by existing body plans, funnel evolution through an increasingly narrow pathway that ends in a singularity. Each step of an evolutionary process restricts the options open to the following step or, otherwise put, increases the degree of *commitment* in a given direction. The difference between this kind of bottleneck and the preceding one is that it is shaped by internal rather than external factors. These, of course, are theoretical extremes. In reality, the two kinds of factors are often involved simultaneously to varying degrees.

Mechanism 4. Pseudo-Bottleneck. This term is used to describe the case in which a single branch emerges, not through some sort of selective or restrictive process, but as a mere consequence of the progressive attrition of all the others. In this form of singularity, historical contingency plays a much more important role than it does in bottlenecks, whether externally or internally imposed. The two forms may, however, not be easily distinguishable from one another. There probably exists a continuous gradation between the chance survival of a given branch and its enforced emergence by a combination of selective and restrictive factors.

Mechanism 5. Frozen Accident. In this case, the singularity is due to pure chance deciding between two or more possibilities that happen to be such that a course can no longer be changed once it has been set. Imagine arriving at a road fork and flipping a coin to decide whether to turn left or right. Or make it a roundabout with six outlets and base your decision on the throw of a die. The main point is that all options are open until you have left it to chance to decide for you. After that, you are irrevocably committed; all other options are henceforth ruled out. Chance decides, but, once the decided direction has been taken, there is no going back and trying another one.

Mechanism 6. Fantastic Luck. In this mechanism, chance plays an even greater role than in the preceding one, in that the singularity is actually attributed to a highly improbable event, most unlikely to be repeated anywhere, any time. There is nothing untoward about such a possibility: Single events of extremely low probability take place all the time in nature and should not excite any wonder. An example I have cited before comes from the game of bridge. Each distribution of the 52 cards among the four players is one out of a total of 5×10^{28} possible distributions. This figure is truly immense. Even if the entire present population of the world had been playing bridge nonstop since the time of the Big Bang, only a small fraction of this number of distributions would have been dealt. Thus, the probability of any given card distribution is infinitesimally small. Yet no bridge player has ever exclaimed at being witness to a near-miracle. What would be a miracle, however, or, rather, unmistakable evidence of trickery, is if the same distribution should be dealt again, even once. The probability of such an occurrence is one in 2.5×10^{57} ($5 \times 10^{28} \times 5 \times 10^{28}$), that is, for all practical purposes, zero. Thus, in the realm of very low probabilities, chance and singularity go together, an argument often invoked in support of uniqueness in evolution.

Mechanism 7. Intelligent Design. This mechanism postulates the occurrence of evolutionary steps that could not possibly have taken place without the intervention of some kind of supernatural guiding entity. Strictly speaking, such a possibility hardly deserves mention in a scientific context, as it can come into account only after all natural explanations have been ruled out, and, obviously, they never can be. Intelligent design has, however, been advocated in recent years by a small minority of highly vocal scientists, who claim to have demonstrated that certain evolutionary steps cannot be explained in strictly natural terms. Loudly acclaimed in many fundamentalist and even more liberal religious circles, these arguments have failed to convince a significant number of scientists.

1 Building Blocks

All living organisms, from microbes to humans, are made of the same basic building blocks, consisting mainly of sugars, amino acids, fatty acids, and nitrogenous bases, altogether little more than fifty distinct, small chemical species, of molecular weights rarely exceeding 200. What largely differentiates organisms chemically is the manner in which these building blocks join into larger assemblages, mostly polysaccharides, proteins, lipids, and nucleic acids. A number of additional compounds peculiar to certain sets of organisms exist – chlorophyll in plants is an example – but these are most likely products of later evolution. In its earliest forms, life probably was made of little more than the universal building blocks found in all living organisms today.

Prebiotic Chemistry

This remarkable singularity goes back to the very beginnings of life, conveying a central message whose meaning did not catch the attention of biologists until 1953, when Stanley Miller observed the spontaneous formation of a number of amino acids and other typical biological constituents in a laboratory situation he and his mentor, the celebrated chemist Harold Urey, believed to be representative of the conditions that prevailed on the primitive Earth at the time life first appeared (Miller, 1953). Even though serious doubts have since been voiced about its underlying assumption, Miller's achievement remains a major landmark in the history of biology. It opened the new field of prebiotic chemistry and sparked a large number of interesting experiments.

Perhaps even more importantly, Miller's findings highlighted the possibility that the building blocks of life could have been the products of natural chemical phenomena, mandated by local physical–chemical conditions. This possibility should not have come as a surprise, for it was consistent with the view, already solidly established at that time, that living processes take place naturally, without the intervention of some special "vital force." But, with rare exceptions, biochemists in those days hardly bothered with the origin of life, which they relegated to the realm of the unknowable, not worth experimenting on or even thinking about.

Cosmic Chemistry

Surprisingly, nothing like the sensation created by Miller's experiments greeted the later, much more startling discoveries showing the presence of a variety of organic molecules in extraterrestrial sites never visited by any living organism. Detected by radioastronomical spectroscopy, by spacecraft-borne instruments sent to passing comets and other parts of the solar system, and by detailed analyses of meteorites that have fallen on Earth, these organic substances now number in the hundreds, creating a host of fascinating problems for the scientists who attempt to explain their formation (for two comprehensive reviews, see Botta and Bada, 2002; Ehrenfreund et al., 2002).

From the ultra-rarefied interstellar spaces, where hours may go by before an atom encounters another atom, to the tiny dust particles that float in space, to the material that makes up protoplanetary disks around new-forming stars, to the comets and asteroids that condense from this material, to the surfaces of planets and their moons orbiting around a star, there exists a gradation of complexity, ranging from small assemblages of a few atoms – such as the hydroxyl, methyl, and methylene radicals, water, methane, carbon monoxide, methanol, formaldehyde, cyanide, cyanate, isocyanate, and thiocyanate, to mention only a few – to a wide variety of larger molecules, including, in addition to a considerable amount of "junk" (as judged by a biochemist), amino acids, sugars, purines, pyrimidines, fatty acids, and other typical biological constituents, some engaged in more complex associations.

These observations have now reached the stage of laboratory experimentation. In France, the group of Guy Ourisson has obtained a wide variety of biological building blocks by high-energy bombardment of graphite targets with molecular beams of the desired atoms, simulating plausible interstellar events (Devienne et al., 1998, 2002). In another approach, two groups, one American (Bernstein et al., 2002) and the other European (Munoz Caro et al., 2002), have found that a number of amino acids arise spontaneously in simulated interstellar ice analogues containing water, methanol, and ammonia as main components, with, in addition, hydrogen cyanide in one case (Bernstein et al., 2002), and carbon monoxide and carbon dioxide in the other (Munoz Caro et al., 2002), exposed to UV irradiation at very low temperature and under very high vacuum.

There has been much discussion concerning the possible destruction of extraterrestrial organic molecules upon entry into the atmosphere, and later impact, of their carrying body, whether comet or meteorite (Botta and Bada, 2002; Ehrenfreund et al., 2002). It is generally believed that, even though such destruction may have been important, a sufficient proportion of the incoming material would have been spared, providing abundant building blocks for the origin of life on our planet. A partial contribution by terrestrial chemistry is, however, not excluded.

The general conclusion that emerges from all these findings is that *the building blocks of life form naturally* in our galaxy and, most likely, also elsewhere in the cosmos. The chemical seeds of life are universal. The first singularity that we detect is thus the consequence of the basic laws that govern the transformations of matter in the universe; it is clearly of *deterministic* origin (mechanism 1).

This conclusion presupposes that the products of cosmic chemistry (and of terrestrial prebiotic chemistry) did indeed serve to initiate the development of life on Earth. The remarkable similarities between these products and the building blocks of life would seem strongly to support this assumption. Nevertheless, as will be mentioned in Chapter 5, a theory vigorously advocated by some investigators holds that life started from scratch, so to speak, that is, from carbon dioxide and other simple materials developing into increasingly complex compounds by pathways that prefigured in some respects the reactions of present-day metabolism. As we shall see, these are very interesting proposals, which include a number of valuable aspects. But the assumption that the products of prebiotic chemistry may have had nothing to do with the initiation of life on Earth – and, therefore, that their similarities with the building blocks of life are a meaningless coincidence – does seem difficult to accept.

Note, however, that not all the products of cosmic chemistry are used for the construction of life. For example, α-aminobutyric acid and α-aminoisobutyric acid, which are present in fair abundance in the Murchison meteorite and in Miller's experimental flasks, are not found in living organisms. On the other hand, ubiquitous compounds, such as the amino acids tryptophan and histidine, have not so far been identified among products of cosmic chemistry and may never be, having appeared later in the development of life. Thus, selection and innovation are two key notions that have to be added to determinism at an early stage in our appreciation of life's building blocks. A remarkable singularity, in this connection, is the fact that life often uses only one of the two possible isomers whenever molecules exhibiting the phenomenon of chirality are employed. Known as homochirality (Greek for "same-handedness"), this trait, which is considered by many as one of the most mysterious properties of life, deserves to be examined in some detail.

2 Homochirality

It has been known since the days of Pasteur (famous, notably, for the separation of two forms of tartaric acid) that molecules containing an asymmetric carbon atom, that is, a carbon atom bearing four different groups, can exist in two forms that are to each other like one hand to the other (*cheir* means hand in Greek) or like an object to its image in a mirror. When aqueous solutions of such molecules are traversed by a beam of polarized light, the polarization plane of the light is rotated by a certain angle. The value of this angle, adjusted to the concentration of the solution and to the thickness of the liquid layer traversed, is known as the specific rotatory power, or optical activity, of the substance; it is the same in absolute value, but of opposite sign, for the two forms. By definition, the optical activity is said to be positive when the polarization plane of the light is rotated to the right, and negative in the opposite case. The two forms, known as enantiomers (*enantios* means opposite in Greek), are designated *d*, for dextrorotatory (*dexter* means right in Latin), and *l*, for levorotatory (*laevus* means left in Latin).

Following a proposal made at the beginning of the last century by the German chemist Emil Fischer, the nomenclature based on optical activity was replaced by one based on structure. Stereoisomers (*stereos* means solid in Greek) replaced optical isomers. Fischer based his classification on the two forms of glyceraldehyde, designating by D the *d* form of this substance, in which the central OH group lies at the right in the planar projection of the molecule (the aldehyde group being written on top), and by L the *l* form, in which this group lies at the left:

$$
\begin{array}{cc}
H-C=O & H-C=O \\
| & | \\
H-C-OH & HO-C-H \\
| & | \\
CH_2OH & CH_2OH
\end{array}
$$

D-glyceraldehyde   L-glyceraldehyde

Fischer classified sugars according to this rule, the prefix D being given to the form in which the OH group attached to the penultimate carbon atom (next to the terminal CH_2OH) lies at the right, and the prefix L being given to the corresponding enantiomer. In this convention, the letters D and L no longer indicate the sign of the optical activity. For example, whereas D-glucose is dextrorotatory, D-fructose, which combines with glucose to form sucrose, the common sugar, is levorotatory. So are D-ribose and D-deoxyribose. Optical rotation, when indicated, is represented by a sign, for example, D-(+)-glucose, D-(−)-ribose, and so on.

This nomenclature was extended to α-amino acids, the position of the α-amino group (the carboxyl group being written on top) serving to distinguish the D and L forms:

$$
\begin{array}{cc}
COOH & COOH \\
| & | \\
H-C-NH_2 & H_2N-C-H \\
| & | \\
R & R
\end{array}
$$

D-amino acid   L-amino acid

As it happens, L-amino acids are levorotatory, at least in neutral solution (they become dextrorotatory in an acidic medium). So, in their case, L corresponds to *l*.

In living organisms, most of the complex substances (polymers) constructed from chiral building blocks (monomers) are homochiral, that is, contain molecules of the same stereochemical handedness. Proteins are made exclusively of L-amino acids (except for glycine, which is not optically active). Among nucleic acids, RNA contains only D-ribose, and

DNA D-deoxyribose. Starch and glycogen are homochiral combinations of D-glucose, and so on. There are exceptions. For example, D-amino acids are found, alongside L-amino acids, in murein, the main constituent of eubacterial cell walls, and in a number of bacterial peptides and related substances; they are even found in significant amounts in higher organisms (Fujii, 2002).

The remarkable homochirality of central biological constituents has been the object of much discussion and speculation. Actually, there are two separate problems, represented by the "homo" and "chirality" parts of the term. The use of building blocks of the same (homo) chirality for the construction of biological macromolecules can be largely explained by the fact that these macromolecules would not display their characteristic properties if they were not so constructed. Proteins made of the two kinds of amino acids could not adopt typical structures, such as α-helices or β-sheets, that are essential to their biological properties. Heterochiral nucleic acids most likely would not be replicable (Schmidt et al., 1997). Thus, whatever the starting situation, one would expect homochirality to emerge by selection. This question will be examined in subsequent chapters.

The key problem raised by biological homochirality concerns the "choice" between enantiomers. It cannot possibly be due to the fact that building blocks were available in only one chiral form for the early life-forming processes. Uncatalyzed chemical syntheses of chiral substances invariably yield racemic mixtures, that is, equal amounts of the two isomers. Among catalysts, enzymes are almost unique in being chirally specific. This property is so remarkable that the artificial creation of chiral catalysts has been considered sufficiently noteworthy to warrant awarding of the 2001 Nobel prize in chemistry to their inventors. The fact that nature sometimes uses the two enantiomers of chiral substances (see above) further suggests that both isomers must have been available originally. It seems almost certain, therefore, that the first building blocks made by cosmic and terrestrial chemistry were racemic. The "choice" must have occurred subsequently.

How this could have happened is not known. A possibility not widely entertained but nevertheless worthy of consideration is that the first catalysts involved in the utilization of the building blocks for assembly reactions were already chirally specific. A related possibility, to be considered in Chapter 8, is that the chirality of RNA may have determined the chirality of the proteins; in other words, the "choice" of D-ribose for the synthesis of RNA dictated the "choice" of L-amino acids for the assembly of proteins. This would simplify matters by reducing the "choice" to the sole ribose molecule, instead of nineteen amino acids, making it a true singularity.

According to many workers in the field, this "choice" could be a typical example of a frozen accident (mechanism 5). Two equally probable alternatives presented themselves. Pure chance decided in favor of one, which, once adopted, could not be reversed. But were the two alternatives equally probable?

This question has attracted the attention of a number of physicists, who have looked into possible mechanisms whereby given enantiomers could be preferentially synthesized or, more likely, destroyed by such phenomena as parity violation or asymmetric irradiation, in particular UV circular dichroism (for a brief review, see Jorissen and Cerf, 2002). The resulting enantiomeric excess, although small, would have been sufficient to tip the scale in favor of the more abundant enantiomer at the time selection occurred.

The analysis of meteorites has revealed a systematic and often considerable excess of L- over D-amino acids (Botta et al., 2002). This fact was generally attributed to inevitable contamination by debris from living organisms until the finding, in 1997, that some amino acids that are not used for protein synthesis may sometimes also show an excess of the L form, though much more modest than that observed for proteinogenic amino acids (Cronin and Pizzarello, 1997). The possibility of a slight cosmic bias in favor of L-amino acids, at least in the neighborhood of the solar system, therefore cannot be excluded. Interestingly, a recent study has shown that such a bias is catalytically communicable: The chirality of four-carbon sugars (tetroses) formed from glycolaldehyde by amino-acid-catalyzed aldol condensation was found to depend on the chirality of the amino acids (Pizzarello and Weber, 2004).

We must leave this matter to experts. Its main relevance, from the point of view of this book, besides its obvious theoretical interest, is to extraterrestrial life. Could there be other life forms containing the same basic constituents as terrestrial organisms, but of opposite chirality? Should such forms one day be discovered, their chirality would be strong evidence of their independent origin.

3 Protometabolism

Given a steady supply of organic building blocks, whether delivered from outer space or formed locally, what goes on next depends on prevailing physical and chemical conditions, as exemplified by Earth, Europa, Titan, and, perhaps, Mars, which all have undergone different histories while probably all receiving the same array of cosmic building blocks. Under the conditions that existed on Earth some four billion years ago – assuming life arose on our planet – the building blocks entered into a complex set of chemical interactions that initiated a long road that finally was to lead to the first primitive living cells. These early chemical processes are generally referred to as prebiotic, or abiotic, chemistry. They will be designated *protometabolism* in this book, in order to indicate that they preceded present-day metabolism and gave rise to it.

The pathways of protometabolism are unknown, but they can be approached, theoretically and experimentally, with the help of what is known of present-day life. There are good reasons to believe, for example, that RNA was the first bearer of replicable genetic information,[1] only

---

[1] The possibility that RNA may itself have been preceded by some simpler, replicable, information-bearing substance has been repeatedly evoked. A whole literature exists on the topic, which has produced a number of findings that are interesting from the chemical point of view but not obviously relevant biologically. Being unsupported by direct evidence, these suggestions will not be considered in this book.

later to be followed and replaced by DNA. It is also highly probable that proteins were originally assembled from amino acids by interacting RNA molecules, which eventually gave rise to the components of the protein-synthesizing machinery. These notions, which have inspired Gilbert's RNA-world model (Gilbert, 1986), will be examined in greater detail later. They are essential for the understanding of the intriguing transition from protometabolism to metabolism and thereby throw light also on the properties of protometabolism itself.

Metabolism and Enzymes

In today's living world, metabolism obeys the laws of chemistry, but it in no way follows the pathways organic chemists would adopt to effect the same transformations. The difference lies in enzymatic catalysis. Each of the many chemical reactions that take place in living cells – several hundred at the very least, several thousand in many instances – is catalyzed by an enzyme. Almost without exception, not one of those reactions would proceed at even a very slow rate without such catalytic support. Organic chemists, not having enzyme-like catalysts at their disposal, must rely on different mechanisms when they try to mimic nature. Their procedures often involve several steps, each carried out in a selected medium and separated from the next one by purification of the product.

Most enzymes are proteins, that is, molecules made of a large number of amino acids strung together in specific sequences dictated by the nucleotide sequences of messenger RNA molecules, which are themselves transcribed from DNA sequences. As mentioned above, proteins were most probably "invented" by RNA, which was also the first bearer of replicable genetic information, before DNA. Thus, the first enzymes were RNA "creations."

A few biological reactions are catalyzed by RNA molecules (ribozymes),[2] rather than by protein enzymes. Even if, as is assumed by the RNA-world hypothesis (Gilbert, 1986), ribozymes fulfilled a more important role in prebiotic times than today, their appearance must, by

---

[2]   As will be seen in later chapters, the two main functions of catalytic RNAs in present-day life are in protein synthesis and RNA processing.

necessity, have followed the chemical development of RNA itself. Hence, all the chemical phenomena that preceded the appearance of RNA and led to it must have taken place without the help of either RNA or protein catalysts, from which it is generally concluded that protometabolism cannot have followed the pathways of metabolism.[3]

Congruence

The reasoning mentioned above seems unassailable, but it meets with two difficulties. One is that the many attempts by chemists to duplicate nature by plausible prebiotic processes have so far met with only limited success, in spite of a considerable expenditure of effort and ingenuity. This may be a question only of the right approach not having been taken. This failure may be corrected tomorrow, and then everyone will exclaim: "Why didn't I think of this before?"

True enough. But there remains another, more fundamental question: How did protometabolism come to be replaced by metabolism? The obvious answer to this question is that the appearance of catalysts, whether ribozymes, protein enzymes, or both, was responsible for the transition, delineating the new pathways by their catalytic activities. But this does not solve the problem. We have to ask *how* catalysts with appropriate properties came to appear. If we rule out intelligent design and, which amounts almost to the same thing,[4] strict determinism, the only scientifically plausible explanation is that the catalysts arose through *selection*. This explanation has profound implications.

Selection is possible only in a system endowed with replicability and variability. The first property presumably belonged to RNA molecules (DNA would serve as well, but the mechanism would have to be more complex, requiring an additional transcription step). The second property,

---

[3] According to Leslie Orgel, a pioneer of origin-of-life research and a forceful defender of this view, metabolism, being a creation of ribozymes and protein enzymes, could not throw any light on the origin of its creators. "If the RNA world developed *de novo*," he writes in a recent paper, "it erected an opaque barrier between biochemistry and the prebiotic chemistry that preceded the RNA world" (Orgel, 2003).

[4] Intelligent design and strict determinism may be equivalent in terms of practical outcome; they are, of course, profoundly different philosophically.

variation, follows automatically from the first, especially in a primitive system where replication must have been very imprecise. There remains selection itself. How could the possession of some catalytic activity by an RNA or protein molecule favor that molecule's selection?

If we are dealing with a ribozyme, the catalytic property could conceivably go together with a greater stability of the molecule, or with its faster replication, or with both, leading to molecular selection of the kind first reproduced in a test tube by Spiegelman (1967). It is, however, doubtful that such a mechanism could account for the selection of catalytic RNAs in general. It could not possibly explain the selection of proteins, which are not replicable. For these substances, selection must perforce have been indirect. The protein enzyme must have exerted a positive feedback on the replication of the RNA coding for it [Eigen's hypercycle (Eigen and Schuster, 1977)]. As a general mechanism, this could have been achieved only by way of competing, reproducing units, or *protocells*, that derived some advantage from the possession of the enzyme (de Duve, 2005).

We are thus led to imagine, as a necessary precondition to the selection of enzymes (and, probably, also many ribozymes), the existence of a large number of protocells, each endowed with a primitive, replicable RNA genome, a rudimentary translation machinery, and the ability to grow, divide, multiply, and thus compete for available resources. How these features appeared will be considered later in this book. Suffice it to state that they must, by definition, have been acquired without the help of enzymes, with the sole support of protometabolism. It follows that protometabolism must have displayed an appreciable degree of *complexity* to be able to support the development of protocells thus endowed, and that it must have enjoyed considerable *stability* and *robustness* under prevailing conditions to do so during the whole time required by this development.

Let us now consider the mechanism of selection itself. According to the proposed Darwinian model, the triggering event must be a genetic modification, the result of a replication error, no doubt plentiful at that time, or of some other accident, leading to the production by the affected protocell of a protein endowed with some catalytic activity. Should this activity prove useful to the mutant protocell, and only then, selection and amplification of this protocell will take place, eventually leading to a population of which most members possess the enzyme. Repeat of the same mechanism for another enzyme will likewise cause the selection and amplification of protocells endowed with the two activities. These protocells may then, in turn, similarly acquire a third enzyme, and so on.

As depicted, the proposed scenario is, of course, a highly schematic and oversimplified view of what may have happened. Some mutations could have taken place simultaneously, rather than sequentially. Protocells could have exchanged genes on a large scale, as happens today among bacteria.[5] More than one population of protocells could have been involved. All kinds of possibilities can be contemplated. What is crucially important and common to all models is the key role of selection and the consequent prerequisite of *usefulness*.

For the protocellular owner of an enzyme to be selected, it must have derived an advantage from this ownership. The enzyme must have been useful. Absent this property, there could have been no selection. Now, it is clear that a catalyst can be useful only if it finds in its surroundings one or more substrates on which to act and one or more outlets for the product(s) of the reaction(s) it catalyzes. Without a substrate, even the most sophisticated catalyst is useless. On the other hand, if what it makes cannot be cleared, the catalytic activity will lead into a blind alley, serving no advantage and possibly even being harmful. Substrates and outlets must have been provided by protometabolism, which thus acted as a screen in the selection of the first enzymes. Only those enzymes that fitted within the existing chemistry were conserved, from which it follows that protometabolism must have prefigured metabolism. The two were *congruent*; they followed similar pathways.[6]

Note that this argument applies also to any ribozyme that could not have been retained by direct molecular selection. Thus, even a hypothetical RNA world supported largely by catalytic RNAs (Gilbert, 1986)

---

[5] We shall see in Chapter 14 that several workers believe horizontal gene transfer to have been very frequent among early protocells.

[6] The congruence argument, which I first made in 1993 (de Duve, 1993) and have reiterated many times since, has not so far been addressed critically by other workers in the field, even though it opens a wide tunnel in the "opaque barrier" alluded to by Orgel (2003) in the paper quoted above (see footnote 3).

could have inherited a significant part of its underlying chemistry from protometabolism.

Theoretical considerations thus lead to apparently paradoxical conclusions. Protometabolism, having had no enzymes to support it, must necessarily have been very different from metabolism, which relies entirely on enzymes. Yet, metabolism could not have arisen historically without taking over many of the pathways or, at least, reactants of protometabolism. Obviously, there must be a flaw somewhere. Let us first look a little more closely at the congruence argument.

One point that I did not clearly state in my earlier writings on the topic and only recently came to appreciate is that the mechanism proposed for the emergence of metabolism is *doubly* selective (de Duve, 2005). Not only are enzymes selected only if they fit into protometabolism, as argued, but reactions of protometabolism are carried over into metabolism only if enzymes catalyzing them arise and are selected. In other words, protometabolism and enzymes are mutually selective. Thus, the appearance of the first enzymes could have signaled an important bottleneck in the development of life, a key transition from what may have been a "dirty" protometabolism to a cleaner metabolism. It is quite possible that the chemistry of emerging life consisted of a set of fairly unspecific reactions, among which appearing enzymes would progressively have selected a coherent subset that turned into metabolism.

The existence of this bottleneck raises an intriguing question: Could metabolism have been different if, by the chance occurrence of other mutations, proteins with different catalytic activities had emerged and a different subset of protometabolic reactions had been selected to form metabolism? In other words, could metabolism be, at least in part, the result of a frozen accident? There is no simple answer to that question. As we have seen, the "choice" of chirality, being a simple binary one, could have been of that kind, although even that is not certain. In the case of metabolism, it is conceivable that the chance appearance of a given enzyme created conditions favoring the subsequent selection of enzymes that would not otherwise have been retained. If, however, conditions had been such as to permit a near-exhaustive exploration of the possibilities offered by chance, *optimization* would eventually have been achieved, whatever the order in which mutations took place.[7] As will be seen in later chapters, the second alternative seems more probable.

One last remark is in order before closing this important question. Congruence with protometabolism applies to the reactions catalyzed by the first, primitive enzymes, not necessarily to the much more complex processes that characterize present-day metabolism. Much may have changed in the course of the long evolutionary sequence of events leading from the early catalysts to the highly sophisticated enzymes of today. This is obviously true. No doubt, much was added to the first reactions as evolution proceeded, and these reactions may, themselves, have undergone considerable modifications. It is, however, unlikely that these changes could have been such as to obliterate completely the legacies of the past. What is known of the modular construction of proteins (see Chapter 8) suggests that certain core catalytic centers going back to early days served in the combinatorial mechanism whereby modern proteins were assembled. To distinguish between true innovation and evolutionary continuity may, however, not be easy.

The First Catalysts

It is widely realized that protometabolism is not likely to have functioned without the help of catalysts. However, most investigators have, for understandable reasons, looked mainly in the mineral world for possible prebiotic catalysts. Their search has not been entirely fruitless. Metal ions and, especially, clays have yielded a number of interesting results, but restricted largely to the single process of RNA assembly.

---

[7] Relevant to this question is an interesting cladistic study by Cunchillos and Lecointre (2002, 2005), who have tried to establish the order in which metabolic pathways have appeared. Their results suggest that metabolism started with the breakdown and synthesis of amino acids. Central pathways, like the Krebs cycle and the glycolytic chain, apparently developed later. It must be noted that the authors of this work believe, contrary to my proposed scheme, that their results apply to *protometabolic* pathways, allegedly catalyzed by authentic proteins taken to have appeared *before* RNA and to have been subject to natural selection without the intervention of nucleic acids. This interpretation, which contradicts notions on the place of proteins in the origin of life that are generally accepted and have been adopted in this book, is not supported by any objective argument.

The congruence notion prompts our taking a new look at the problem. If protometabolism and metabolism were congruent, then the equivalents of a number of key enzymes must have been present in the prebiotic setting. At first sight, this hardly looks plausible. One doesn't see how the barren, prebiotic milieu could possibly have generated substances capable of mimicking the catalytic properties of a number of complex protein molecules. There is a possibility, however.

Together with the French investigator André Brack (2003), I have proposed that prebiotic catalysis could have been accomplished by peptides or, more precisely, by what I have called *multimers* to imply that they could have been made from additional components besides amino acids. Compared to proteins, my hypothetical multimers would have been much shorter, and they would have been more heterogeneous, containing a variety of amino acids of both D and L chiral configuration and, possibly, other constituents, such as hydroxy acids. Interestingly, many substances obeying this description exist in nature. Grouped into two categories, peptides and polyketides, they are made mostly by bacteria or by fungi, and are attracting increasing interest because they include many antibiotics and other active agents, such as the immunosuppressor cyclosporin (Cane, 1997; Walsh, 2004). They are synthesized by a variety of complex mechanisms that have in common that they all use the thioesters (see Chapter 6) of the constituent acids as activated precursors.

The possibility that similar substances could have arisen spontaneously in a prebiotic setting is plausible. Made of building blocks that are abundant among the primary products of cosmic chemistry, the posited multimers could have formed by fairly simple assembly mechanisms, which could even, like their natural counterparts, have involved thioester precursors. It is even possible that substances resembling my multimers may be formed by cosmic chemistry. There is evidence for the presence of peptides in meteorites, though apparently limited to the simple dipeptide glycylglycine (Shimoyama and Ogasawara, 2002). In the simulation experiments reported above (Bernstein et al., 2002, Munoz-Caro et al., 2002), free amino acids were obtained in substantial amounts in the products of UV-irradiated interstellar ice analogues only if the material had first been subjected to acid hydrolysis. The possibility that the amino acids may be present in the form of peptides is explicitly raised by the investigators. It is thus conceivable that peptides and analogous substances accompanied basic building blocks in the cometary showers that are believed to have delivered chemical seeds of life to the prebiotic Earth.

It is also possible that such substances formed later on Earth. Several investigators have described the synthesis of peptides under plausible prebiotic conditions (see, for example, Huber and Wächtershäuser, 1998; Leman et al., 2004). Particularly interesting are the experiments by a group of Japanese workers who have observed amino acid oligomerization in a hot–cold flow reactor designed to simulate a hydrothermal vent setting (Imai et al., 1999; Ogata et al., 2000; Yokoyama et al., 2003; Ozawa et al., 2004).

Whether compounds made in this way could have displayed, be it only in a rudimentary fashion, the catalytic activities required for a metabolism-like prebiotic chemistry has not yet been ascertained. No catalytic activity has been detected in any of the natural substances mentioned above. But this could be because none has been looked for or because no selective pressure favored the emergence of catalysts in the presence of the much more efficient protein enzymes. Being closely related to proteins, the hypothetical multimers certainly offer a good chance of including molecular configurations resembling the active sites of enzymes. Their shortness does not preclude such a possibility.[8] As will be pointed out later (see Chapter 8), there are strong reasons to believe that the first protein enzymes were little more than some twenty amino acids long, on average. If such was the case, it follows that peptides of such small size could display the catalytic activities required by protometabolism. Needless to say, mineral catalysts, in particular metal ions, could have been involved as well, as they are in many enzyme-catalyzed reactions today.

It has been objected that the multimer hypothesis does not include a mechanism for the selection of the active catalysts. But this is true of all the processes that preceded the emergence of the first replicative machinery conducive to natural selection. Until this happened, chemical

---

[8] Even single amino acids can display catalytic activities (Bar-Nun et al., 1994; Weber, 2001; Pizzarello and Weber, 2004).

determinism had to suffice, without the help of selection.[9] The assumption, therefore, is that the multimer mixture that arose spontaneously under the prevailing conditions contained an adequate set of catalysts and that the conditions remained stable long enough to ensure an uninterrupted supply of these catalysts until they could be dispensed with. The first requisite is obviously the decisive one, because stability of conditions would be necessary for the continuing occurrence of any critical chemical step dependent on those conditions, whatever its nature. Whether a set of catalysts sufficient for a metabolism-like protometabolism could possibly have been present in the postulated multimer mixture is a matter of mere conjecture at the present time.

It need not remain so indefinitely. The multimer theory or, at least, its verisimilitude can be put to experimental test. As proposed elsewhere (de Duve, 2003), natural and artificial analogues of my hypothetical multimers could be screened for some key catalytic activities, mainly electron transfers and group transfers. Perhaps such experiments will be carried out some day.

[9] The possibility must be admitted that the composition of the hypothetical multimers was not entirely ruled by strict determinism. In their experiments showing the formation of peptides in a simulated hydrothermal environment, the Japanese workers have noted the occurrence of what they call "stochastic aspects" (Yokoyama et al., 2003).

4 ATP

Throughout the living world, energy circulates almost entirely in the form of a single chemical currency, known as ATP in biochemical jargon. The central function of this substance stands out as one of the most remarkable singularities in the organization of life, all the more impressive because ATP, together with its close relatives and occasional substitutes, GTP, CTP, and UTP, also represents one of the four universal precursors in the construction of RNA, most likely the first information-bearing molecule in the development of life.

Anatomy of a Molecule

ATP stands for adenosine triphosphate. As we shall see in greater detail later, adenosine (A) belongs to the group of *nucleosides*; it is a combination of the purine base adenine with the five-carbon sugar ribose.

Attachment of a phosphate molecule to the ribose end of A by an ester linkage produces adenosine monophosphate (AMP), a *nucleotide*. Two additional phosphoryl groups attached to the terminal phosphate of AMP and to each other by *pyrophosphate* bonds (pyrophosphate is the product of the heat-induced combination, with loss of water, of two molecules of inorganic phosphate) lead successively to adenosine diphosphate

(ADP) and adenosine triphosphate (ATP). These structures are illustrated below:

$$
\begin{array}{c}
\quad\quad O \\
\quad\quad \| \\
A\text{-}O\text{-}P\text{-}O^- \\
\quad\quad | \\
\quad\quad O^-
\end{array}
\quad\quad \text{AMP}
$$

$$
\begin{array}{c}
\quad\quad O \quad\ O \\
\quad\quad \| \quad\ \| \\
A\text{-}O\text{-}P\text{-}O\text{-}P\text{-}O^- \\
\quad\quad | \quad\ | \\
\quad\quad O^- \ \ O^-
\end{array}
\quad\quad \text{ADP}
$$

$$
\begin{array}{c}
\quad\quad O \quad\ O \quad\ O \\
\quad\quad \| \quad\ \| \quad\ \| \\
A\text{-}O\text{-}P\text{-}O\text{-}P\text{-}O\text{-}P\text{-}O^- \\
\quad\quad | \quad\ | \quad\ | \\
\quad\quad O^- \ \ O^- \ \ O^-
\end{array}
\quad \text{ATP}
$$

It is useful to mention at this stage that identical structures exist with A replaced by G (guanosine), C (cytidine), or U (uridine), which are nucleosides in which ribose is combined with the purine base guanine, the pyrimidine base cytosine, or the pyrimidine base uracil, respectively. As will be seen later, these compounds are key precursors in the synthesis of RNA. Furthermore, replacement of the ribose in these molecules by deoxyribose (and of U by T, thymidine) yields similar derivatives, involved in the construction of DNA. From the point of view of this chapter, these differences are immaterial (they are all-important as regards information). The business end of the molecules with respect to energy is represented by the triphosphate part, which is the same and possesses the same energetic properties in all varieties.

Life's Jack of All Trades

Most of the work accomplished by living organisms on our planet is powered by hydrolysis (splitting with the help of water) of the terminal pyrophosphate bond of ATP, giving adenosine diphosphate (ADP) and inorganic phosphate (P_i):

$$A\text{-}O\text{-}\overset{\displaystyle O}{\underset{\displaystyle O^-}{\overset{\|}{P}}}\text{-}O\text{-}\overset{\displaystyle O}{\underset{\displaystyle O^-}{\overset{\|}{P}}}\text{-}O\text{-}\overset{\displaystyle O}{\underset{\displaystyle O^-}{\overset{\|}{P}}}\text{-}O^- + H_2O \longrightarrow A\text{-}O\text{-}\overset{\displaystyle O}{\underset{\displaystyle O^-}{\overset{\|}{P}}}\text{-}O\text{-}\overset{\displaystyle O}{\underset{\displaystyle O^-}{\overset{\|}{P}}}\text{-}O^- + HO\text{-}\overset{\displaystyle O}{\underset{\displaystyle O^-}{\overset{\|}{P}}}\text{-}O^- + H^+$$

In biochemical shorthand,

$$ATP + H_2O \longrightarrow ADP + P_i \qquad (1)$$

This reaction releases *energy*, a term used here and elsewhere in this book as a simpler substitute for the more precise term *free energy*, which combines internal energy and entropy to express the maximum amount of work that can be supported by a chemical reaction, the true quantity of interest. Under conditions approaching those that prevail in living cells, the amount of energy released by the hydrolysis of the pyrophosphate bonds of ATP is on the order of 14 kcal (kilocalories), or 59 kJ (kilojoules), per gram-molecule of substance split. This value is the size of the energy packages used in most biological energy transactions, what may be seen as the standard bioenergetic currency unit. If ATP hydrolysis takes place freely, the energy released by the reaction is dissipated as heat. Living systems contain a number of *transducers* that convert the energy released by the hydrolysis of the terminal pyrophosphate bond of ATP [reaction (1)] into various forms of work.

In animal muscles, for example, the transducers are special protein systems that split ATP but can do so only if they shorten at the same time, even against a certain amount of resistance. Thus, the energy released by ATP splitting is converted into the mechanical work of overcoming a resistance. In cell membranes, special transducers, called active transport systems, or pumps, use the energy released by ATP splitting to force substances into cells or out of them against a concentration gradient, that is, from a site where the concentration of the substances is lower to one where it is higher, thus performing osmotic work. If the substances actively transported in this manner are electrically charged (ions), their transport creates charge disparities, or membrane potentials, the source of electrical work in the nervous system, for example, or in the electric organs of certain fish. In luminescent animals, such as fireflies, light is likewise produced at the expense of energy released by the splitting of ATP (with, in this case, the additional support of an oxidative reaction). Particularly important, the innumerable forms of chemical work involved in the biosynthesis of natural compounds and their associations are almost invariably fueled by the splitting of ATP.

Most forms of work appeared late in the development of life, when cells had already attained a high degree of complexity. An exception concerns chemical work, which obviously was carried out right from the start. Leaving out the actual synthesis of building blocks, which, as we have seen, could have been supported by whatever energy sources were available in outer space, the main form of chemical work accomplished by incipient life consisted of joining building blocks together into more complex molecules. Such processes almost invariably involve the removal of a hydrogen atom (H) from one reactant and that of a hydroxyl group (OH) from the other, with formation of a water molecule (H_2O). The vast majority of cell constituents, including proteins, nucleic acids, lipids, polysaccharides, and many other complex molecules, are synthesized in this way from small molecules, such as amino acids, sugars, nitrogenous bases, fatty acids, etc. Called *dehydrating condensations*, such assemblies cannot take place spontaneously in aqueous media because the abundance of water thermodynamically favors the opposite reaction of hydrolysis.

In present-day life, this problem is solved by coupling the energy-requiring dehydrating condensations to the energy-yielding hydrolysis of ATP. The reactions involved in these processes are as varied as the molecules they serve to assemble; they fill a good part of biochemistry textbooks. Fortunately, this diversity reduces to a small number of mechanisms, which are themselves applications of a unifying principle of staggering simplicity that is readily explained and provides the key to the understanding of individual processes. This principle is *sequential group transfer*. The pages that follow, summarized from one of my previous works (de Duve, 1984), should make clear what it is all about.

Group Transfer: The Key to Biosynthesis

Consider the following problem. You wish to carry out the energy-requiring dehydrating condensation of X-OH with Y-H, to give X-Y, in an aqueous medium:

$$X\text{-}OH + Y\text{-}H \longrightarrow X\text{-}Y + H_2O \qquad (2)$$

using as source of energy the hydrolysis of ATP into ADP and inorganic phosphate [reaction (1)]. Letting the two reactions run side by side will not work. The energy released by reaction (1) will be dissipated as useless heat, with no power to drive X–OH and Y-H to join and add more water to a medium where it is already overabundant. Life's solution of the problem is as elegant as it is simple. Take ATP and, instead of transferring its terminal phosphoryl group to water, as in reaction (1), transfer this group to the X–OH reactant, in a reaction called *phosphorylation*, leaving ADP, as in reaction (1), but forming the X-yl phosphate derivative instead of inorganic phosphate:

$$
\underset{\substack{\;\;\;\;| \;\;\;\;| \;\;\;\;|\\O^- \; O^- \; O^-}}{A\text{-}O\overset{O}{\underset{}{\overset{\|}{P}}}\text{-}O\text{-}\overset{O}{\overset{\|}{P}}\text{-}O\text{-}\overset{O}{\overset{\|}{P}}\text{-}O^-} + X\text{-}OH \longrightarrow A\text{-}O\overset{O}{\overset{\|}{P}}\text{-}O\text{-}\overset{O}{\overset{\|}{P}}\text{-}O^- + X\text{-}O\overset{O}{\overset{\|}{P}}\text{-}O^- + H^+
$$

or, in biochemical shorthand,

$$ATP + X\text{-}OH \longrightarrow ADP + X\text{-}O\text{-}P \qquad (3)$$

Now, in a second step, let the X-yl group be transferred from the X–O–P intermediate to the Y–H reactant:

$$X\text{-}O\overset{O}{\underset{O^-}{\overset{\|}{P}}}\text{-}O^- + Y\text{-}H \longrightarrow X\text{-}Y + HO\overset{O}{\underset{O^-}{\overset{\|}{P}}}\text{-}O^-$$

In biochemical shorthand,

$$X\text{-}O\text{-}P + Y\text{-}H \longrightarrow X\text{-}Y + P_i \qquad (4)$$

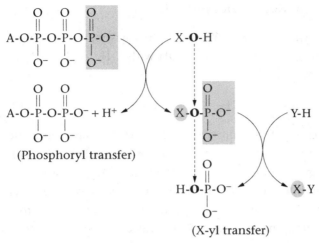

(X-yl transfer)

Figure 4.1. *Sequential group transfer dependent on phosphoryl transfer.* In the first step, the terminal phosphoryl group of ATP is transferred to the X–OH reactant, leaving ADP and forming X-yl phosphate, from which the X-yl group is then transferred to the Y–H reactant in the second step, leaving inorganic phosphate. Note the double-headed intermediate, X-yl phosphate, consisting of the two groups transferred in the two steps, joined by an oxygen atom. This oxygen atom originates from X–OH and ends up in inorganic phosphate, thus tracing the cryptic pathway of water from the dehydrating condensation of X–Y to the hydrolysis of the terminal pyrophosphate bond of ATP.

Add reactions (*3*) and (*4*), and you get

$$\text{ATP} + \text{X-OH} + \text{Y-H} \longrightarrow \text{X-Y} + \text{ADP} + \text{P}_i \tag{5}$$

X–Y is assembled and ATP is split, but water nowhere appears in free form. It travels cryptically with the oxygen of X–OH to the oxygen of inorganic phosphate by way of the X–O–P intermediate. This is seen more clearly in the representation of Figure 4.1.

The essence of this mechanism is that it allows the energy released by ATP hydrolysis to be used for the assembly of X–Y. Instead of being dissipated, as it is in straightforward hydrolysis [reaction (*1*)], this energy moves with the phosphoryl group transferred in the phosphorylation step [reaction (*3*)] and is conserved in the X–O–P intermediate, with the result that the X–OH reactant is now *activated*, that is, fitted with the energy needed for joining with the Y–H reactant [reaction (*4*)].[1] I call the X–O–P

---

[1] Organic chemists similarly use activation – of carboxylic acids as anhydrides, for example – for powering syntheses, but their terminology is different. They generally

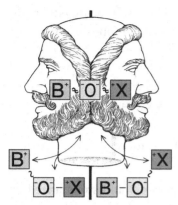

Figure 4.2. *Janus, the double-headed intermediate.* This cartoon represents the obligatory intermediate in all sequential group transfer processes. B stands for the piece of ATP transferred to the X–OH reactant in the initial activation reaction (phosphoryl, AMP-yl, or pyrophosphoryl group). The intermediate is designated "double-headed" because it is composed of two transferable groups linked by an oxygen atom: on one side, the B-yl group, and on the other, the X-yl group, which is transferred in the second biosynthetic step. Note that, thanks to the double-headed intermediate, the oxygen atom provided by the X–OH reactant leaves with the B piece donated by ATP (inorganic phosphate, AMP, or inorganic pyrophosphate), thus coupling dehydrating condensation with ATP hydrolysis. (From de Duve, 1984, p. 125.)

intermediate *double-headed* because it consists of the two transferred groups, linked by the crucial central oxygen atom, which moves, thanks to it, from X–OH to inorganic phosphate (Figure 4.2).

Biochemists will recognize in the mechanism just outlined a number of processes in which the X–OH reactant is a carboxylic acid, which is activated by phosphorylation to the corresponding acyl phosphate. This complex then donates the acyl group to an acceptor that may be ammonia (forming an amide, e.g., asparagine or glutamine), an amine group (in the synthesis of glutathione, for example), or a thiol (making a thioester; see Figures 6.1 and 6.2, below.) Carboxylation (see below Figure 10.2) is another example.

On the whole, the mechanism based on phosphorylation is used relatively infrequently in nature. The most important biosynthetic processes depend on the splitting of the *inner* pyrophosphate bond of

describe transfer reactions as "attacks" (mostly nucleophilic) by the group acceptor on the group donor. This terminology is more rigorous but less intuitive than the notion of group transfer used in biochemistry.

ATP, the group transferred in the first (activation) step being either the AMP-yl group (*nucleotidyl transfer*) or the terminal pyrophosphoryl group (*pyrophosphorylation*):

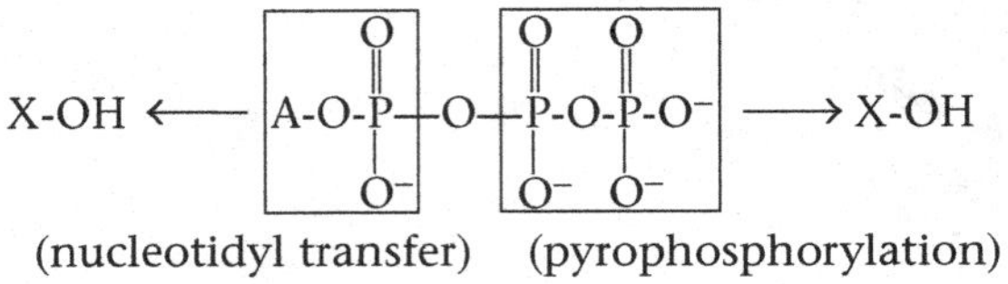

(nucleotidyl transfer)   (pyrophosphorylation)

The first mechanism is illustrated in Figure 4.3 and summarized in biochemical shorthand as follows:

$$\text{ATP} + \text{X-OH} \longrightarrow \text{PP}_i + \text{X-O-AMP} \tag{6}$$

$$\text{X-O-AMP} + \text{Y-H} \longrightarrow \text{X-Y} + \text{AMP} \tag{7}$$

$$\overline{\text{ATP} + \text{X-OH} + \text{Y-H} \longrightarrow \text{X-Y} + \text{AMP} + \text{PP}_i} \tag{8}$$

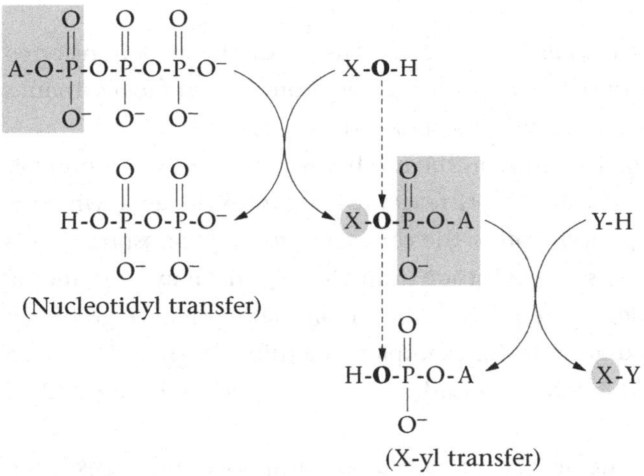

(X-yl transfer)

Figure 4.3. *Sequential group transfer dependent on nucleotidyl transfer.* In the first step, the nucleotidyl group of ATP is transferred to X–OH, leaving inorganic pyrophosphate and forming X-yl AMP, from which the X-yl group is then transferred to the Y–H reactant in the second step, leaving AMP. X-yl AMP is the double-headed intermediate. Its central oxygen atom comes from X–OH and ends up in AMP.

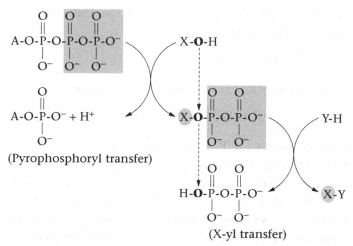

(Pyrophosphoryl transfer)

(X-yl transfer)

Figure 4.4. *Sequential group transfer dependent on pyrophosphoryl transfer.* In the first step, the terminal pyrophosphoryl group of ATP is transferred to X–OH, leaving AMP and forming X–yl pyrophosphate, from which the X–yl group is then transferred to the Y–H reactant in the second step, leaving inorganic pyrophosphate. X-yl pyrophosphate is the double-headed intermediate. Its central oxygen atom comes from X–OH and ends up in inorganic pyrophosphate.

The mechanism dependent on pyrophosphorylation is shown in Figure 4.4. In shorthand,

$$\text{ATP} + \text{X-OH} \longrightarrow \text{AMP} + \text{X-O-PP} \tag{9}$$

$$\frac{\text{X-O-PP} + \text{Y-H} \longrightarrow \text{X-Y} + \text{PP}_i}{} \tag{10}$$

$$\text{ATP} + \text{X-OH} + \text{Y-H} \longrightarrow \text{X-Y} + \text{AMP} + \text{PP}_i \tag{11}$$

Several remarks are in order. It will be noted first that the two processes show the same key features as the mechanism by phosphorylation. Water is nowhere to be seen. The oxygen of X–OH moves to a final ATP split product (either AMP or inorganic pyrophosphate) by way of the double-headed intermediate, which is X–O–AMP in the first case and X–O–PP in the other. The principle of sequential group transfer is universal.

It must be noted next that the final outcome is the same whether the nucleotidyl or the pyrophosphoryl group is transferred [reactions (8) and (11) are identical]. In both cases, X–Y is assembled at the expense of the splitting of ATP into AMP and inorganic pyrophosphate.

Third, when either of these mechanisms is used, the AMP formed is converted to ADP by phosphorylation from ATP:

$$AMP + ATP \longrightarrow 2\,ADP \qquad (12)$$

and the pyrophosphate is hydrolyzed to inorganic phosphate:

$$PP_i + H_2O \longrightarrow 2\,P_i \qquad (13)$$

The end result is that two ATP molecules, instead of one, are hydrolyzed to ADP and inorganic phosphate. The energy expenditure is twice that of the mechanism dependent on phosphorylation, allowing reactions that are particularly costly in terms of energy to proceed readily.

A fourth point is that it sometimes happens that a nucleoside triphosphate (NTP) other than ATP serves as energy donor. The energetics, as we have seen, are exactly the same. When this happens, the NTP used is regenerated from its breakdown product, either NMP or NDP (N stands for nucleoside), by the transfer of terminal phosphoryl groups from ATP, which thus still ends up footing the energy bill:

$$NMP + ATP \longrightarrow ADP + NDP \qquad (14)$$

$$NDP + ATP \longrightarrow ADP + NTP \qquad (15)$$

Finally, the second step (X–yl transfer), especially in the process dependent on nucleotidyl transfer [reaction (7)], is sometimes divided into two successive reactions by the intercalation of a group carrier:

$$X\text{-}O\text{-}AMP + Carrier \longrightarrow AMP + Carrier\text{-}X \qquad (16)$$

$$\underline{Carrier\text{-}X + Y\text{-}H \longrightarrow X\text{-}Y + Carrier \qquad (17)}$$

$$X\text{-}O\text{-}AMP + Y\text{-}H \longrightarrow X\text{-}Y + AMP \qquad (18)$$

Reaction (18), it will be noted, is identical to reaction (7).

The sequence dependent on nucleotidyl transfer is by far the most important biosynthetic mechanism. It serves in the activation of fatty acids for lipid synthesis, with coenzyme A as carrier [see reaction (15), Chapter 6] and in that of amino acids for protein synthesis, with transfer RNA as carrier [see reaction (4), Chapter 8]. It is involved, with CTP replacing ATP, in the synthesis of many phospholipids, and, most often with UTP, in the assembly of polysaccharides. In the latter case, however, the X–OH reactant is a sugar phosphate of which only the sugar group is transferred, while the phosphate group accompanies the leaving UMP, which thus appears as UDP.

A particularly important variant of the mechanism operates in the synthesis of RNA, in which the X–OH reactant is the growing RNA chain and donation of the NMP group (from any one of the four NTPs (ATP, GTP, CTP, or UTP) in reaction (6) actually completes the synthesis (see below, Figure 7.2). The same process serves in DNA synthesis, with deoxyribose derivatives instead of ribose derivatives.

The mechanism dependent on pyrophosphorylation is used infrequently. Its two main applications are in the synthesis of nucleotides and in that of terpenes (with, as activated intermediate, phosphoribosyl pyrophosphate in the first case and isopentenyl pyrophosphate[2] in the second. (See below, Figure 10.3).

The beauty of the biological mechanisms just outlined lies in their centralization. A single reaction, the splitting of ATP to ADP and inorganic phosphate, ends up supporting any number of biosynthetic assembly processes. And this is not all. In addition to dehydrating condensations, a number of other metabolic reactions are likewise provided with energy by special mechanisms involving the splitting of one of the pyrophosphate bonds of ATP. As we shall see later, certain reactions known as biosynthetic reductions may also be fueled by ATP.

Other functions, besides straightforward energy transfers, are supported by ATP. Thus, it happens fairly frequently that substances undergo phosphorylation by ATP before entering metabolism. This process provides the substances with an electrically charged "handle" (the phosphate group is electronegative) that may help in positioning them correctly on the enzymes that act on them. Cellular signaling, which, especially in higher organisms, plays an extremely important role in regulation, also depends in a number of different ways on the splitting of ATP (or of GTP). Remembering that ATP splitting also fuels most other energy

---

[2] The pyrophosphoryl group of isopentenyl pyrophosphate is not acquired by pyrophosphoryl transfer, but by two successive phosphoryl transfer reactions.

transducers, we arrive at the conclusion that, with a few minor exceptions, which will be mentioned later, all biological energy expenditures are, in the last analysis, supported by the hydrolysis of ATP to ADP and inorganic phosphate, one of the most noteworthy singularities offered by life today.

ATP is not stored in sizable amounts by living cells and cannot be obtained from the environment. Thus, cells must continually regenerate the ATP they spend to pay their energy bills, by recombining ADP with inorganic phosphate. This process requires energy (a minimum of 14 kcal, or 59 kJ, per gram-molecule, as we have seen). As will be explained in the next two chapters, a small number of specialized reactions couple ATP assembly either to the use of light energy or with some energy-yielding physical or chemical process. But before getting to that, we must address the key question: Why ATP? The question may be divided into two: Why P? Why A? What caused those two substances to be singled out to become such central participants in the development of life?

Why Phosphate?

In 1987, the American biochemist Frank Westheimer published a widely noted article entitled "Why nature chose phosphates" (Westheimer, 1987). In it, he detailed in perceptive fashion the many properties that make the phosphate molecule uniquely adapted to its numerous biological functions, not only in energy transfers, but also in the construction of such central substances as nucleic acids and phospholipids and in that of many important metabolic intermediates. But Westheimer did not address the question *"How* did nature choose phosphates?"* Unless one believes in intelligent design, fitness does not account for use, except through a process of selective optimization. But phosphate must have entered metabolism before replication and its correlates, mutation and selection, came on the scene, presumably with RNA. There must be a chemical explanation for nature's choice of phosphates. As I have discussed elsewhere (de Duve, 1991, 2001), this explanation is far from obvious.

It cannot be a question of abundance. Phosphate is not a very common mineral constituent on Earth, and most of it is tied up in insoluble combinations with calcium, such as apatite. The phosphate concentration in the oceans and freshwaters is very low, on the order of 0.3 μM or lower, which amounts to some 30 μg/L or less. This scarcity often makes the phosphate concentration a limiting factor in the proliferation of microorganisms, as was revealed by the dramatic increase in this proliferation, or eutrophication, observed in many bodies of water after phosphate was added to cleansing agents (now prohibited for this reason).

The rarity of phosphate is such that certain workers have even suggested that it may not have played any role in early prebiotic chemistry (Keefe and Miller, 1995). However, things could have been different before the appearance of life, which may have been largely responsible for the formation of apatite (Gedulin and Arrhenius, 1994). Also, the phosphate content of prebiotic waters could have been higher than it is today if those waters were more acidic (de Duve, 1991).

A possible clue is provided by the fact that some of the functions of ATP are carried out by inorganic pyrophosphate in certain present-day organisms. This finding has prompted the proposal, vigorously defended by the Swedish couple Margaret and Herrick Baltscheffsky (1992), that, in the origin of life, ATP was preceded by inorganic pyrophosphate as provider of usable pyrophosphate bonds. Another entity that could have played such a role in nascent life is polyphosphate, which is composed of many phosphate molecules linked by pyrophosphate bonds. It is present in all forms of life, acts as a reserve of pyrophosphate bonds, and probably also has regulatory functions (Kulaev, 1979; Kornberg et al., 1989). Both pyrophosphate (*pyr* means fire in Greek) and polyphosphates have been detected in volcanic emissions (Yamagata et al., 1991) and could have served as a source of energy for early life. But they are very rare in the present-day world, suggesting that life may have started in a volcanic setting. I will come back to this point later.

An interesting contribution to the problem has been made by the Swedish–American chemist Gustaf Arrhenius and his coworkers, who have observed that protonated phosphate salts, as opposed to the unprotonated apatite, readily condense to pyrophosphate and higher-chain-length polyphosphate when heated at temperatures as low as 250°C

(Arrhenius et al., 1993; Gedulin and Arrhenius, 1994). They have further found that certain layered minerals, such as hydrotalcite, a mixed hydroxide of magnesium and aluminum, have the ability to soak up phosphate compounds from very dilute solutions and to concentrate them up to supersaturated levels in the aqueous spaces that separate the mineral layers. In this way, the minerals can catalyze certain reactions involving phosphate compounds, including the phosphorylation of organic molecules (Pitsch et al., 1995; Kolb et al., 1997).

Whatever the explanation, nature's "choice" of phosphates can hardly have been a gift of chance. It must have been a consequence of the physical–chemical conditions that made up the cradle of life. This requisite puts a severe limit on those conditions and should serve as a guide in our attempts to identify them.

Why Adenosine?

As already mentioned, adenosine, denoted A, is a nucleoside. It consists of two components: a nitrogenous substance, or base, called adenine, and a five-carbon sugar, or pentose, called ribose. The phosphate of AMP is attached to the ribose moiety of the molecule. Other nitrogenous bases similarly combine with ribose to form G, C, and U, which replace A in certain reactions and, together with it, serve in the formation of RNA. As will be discussed in Chapter 7, the prebiotic appearance of these bases can probably be explained by relatively simple reactions involving hydrogen cyanide (HCN), a typical product of cosmic chemistry, as major reactant.

Ribose raises a thornier problem, not so much the synthesis of this sugar, which can take place from as simple a substance as formaldehyde (H_2CO, also a typical product of cosmic chemistry), but its separation from the many other sugars that form at the same time and, as already mentioned (see Chapter 2), the question of chirality. Ribose is a chiral molecule. The stereoisomer of the sugar in adenosine is the levorotatory D-ribose.

The selective biological use of D-ribose has been the object of much work and speculation, largely because of the overriding interest of investigators in the synthesis of RNA. Preferential formation of this sugar has been considered, and some progress has been made in the identification of a chemical mechanism that could have led to this (Eschenmoser, 1999). Alternatively, there is the possibility that, for some hidden chemical reason, D-ribose was preferentially utilized in the assembly of adenosine and other nucleosides. Of interest in this respect is the observation that borate, probably an abundant component of prebiotic waters, exerts a strong stabilizing effect on ribose (Prieur, 2001; Ricardo et al., 2004) and even catalyzes the combination of this sugar with adenine, forming adenosine, in an aqueous medium (Prieur, 2001).

Whether effects of this sort suffice to explain the selective use of D-ribose remains to be seen. Living organisms contain many different sugars, including three other pentoses in addition to ribose, namely arabinose, xylose, and lyxose. It could, therefore, be that the early chemistry gave rise to a great variety of sugars and that D-ribose acquired its unique function for reasons that had little to do with its relative abundance or special qualities. It is significant, in this respect, that the four pentoses formed in equal amounts and were similarly stabilized by borate when glyceraldehyde was allowed to react with glycolaldehyde at alkaline pH in the presence of calcium ions (Ricardo et al., 2004).

It could even be that other sugars besides ribose joined with adenine and with other bases. As already pointed out, one hardly sees how the kind of crude chemistry that must have been operative in prebiotic days could have been so specific as to produce just those substances that are found to have a function in present-day life. It seems more likely that life started with a *gemisch* of many different molecules of similar structure and that those that are found in living organisms emerged by subsequent selection. This key notion has already been mentioned in connection with the selection of enzymes (see Chapter 3). I shall return to it several times.

However complex the mixture of sugars and of bases incipient life had available, an explanation must be found for the joining of sugars with bases. We have seen that borate could have played a catalytic role in this reaction (Prieur, 2001). An involvement of the pyrophosphate bond is also conceivable. A remarkable feature of present-day metabolism is represented by the close interrelationships between sugars and phosphate.

With rare exceptions, sugars such as glucose, fructose, and galactose enter metabolism via a phosphorylation step supported by ATP. Their subsequent processing by way of metabolic pathways such as the glycolytic chain most often involves phosphorylated intermediates. Also, the association of sugars with one another or with other substances, including adenine and other bases, invariably uses molecules that are activated by combination with phosphate, pyrophosphate, or some other phosphate compound.

These facts suggest the existence of an early cauldron in which sugars interacted with pyrophosphate compounds and with nitrogenous bases to produce a variety of sugar phosphates, nucleosides, and more complex derivatives, including ATP and its analogues. The substances that ended up playing an important metabolic role – in particular serving in the synthesis of RNA – emerged from this mixture by selection, as already suggested. Identifying the conditions under which such reactions could have taken place with the sole help of prebiotically available catalysts (multimers?) raises one of the most challenging problems for future research.

5 Electrons and Protons

ATP, the subject of the preceding chapter, is often called the "fuel of life." This is a misnomer. ATP is no more than an intermediary between the provision and the use of energy. What really fuels life is the *fall of electrons* from a higher to a lower energy level. The mechanisms underlying such energy-yielding electron falls and the associated assembly of ATP represent the heart of bioenergetics. Only their essential lineaments will be outlined in this chapter. More details can be found in standard textbooks and in previous publications (de Duve, 1984, 2001).

Energetics of Electron Transfer

Almost invariably, the assembly of ATP from ADP and inorganic phosphate is coupled to the transfer of one or two electrons between two substances, called the electron donor and acceptor for obvious reasons. In such transfers, the donor passes from the *reduced* (electron-rich) state to the *oxidized* (electron-poor) state, while the acceptor passes from the oxidized to the reduced state. In the reverse reaction, the reduced acceptor becomes the donor, and the oxidized donor the acceptor. This explains why such electron transfers are called oxidation–reduction reactions (redox reactions for short), and the substances involved in them are called redox couples because of their ability to exist in both reduced and oxidized form.

It often happens that protons (hydrogen ions, H^+) accompany the transferred electrons, so that hydrogen atoms ($H = H^+ +$ electron), rather than naked electrons, are transferred (transhydrogenation). Sometimes, hydrogen atoms are donated and only electrons accepted, or vice versa. In the former case, the protons left over by the transfer are released into the surrounding water, rendering the medium more acid. In the opposite case, the required protons are taken from the surrounding water and the medium becomes more alkaline.

Redox couples, generally represented as reduced form/oxidized form (for example, Fe^{2+}/Fe^{3+}, $H_2/2\,H^+$, or lactate/pyruvate), are characterized by their *redox potential*, which is a measure, expressed in volts, of the tendency of the reduced form of the couple to release electrons or, reversing sign, of the avidity of the oxidized form for electrons. By convention, redox potentials are measured against the $H_2/2\,H^+$ couple, taken by definition to have a redox potential of zero under standard conditions. Redox couples of higher reducing power, that is, that spontaneously donate electrons to this hydrogen standard, have a negative redox potential. The opposite is true of couples of greater oxidizing power, which spontaneously accept electrons from the hydrogen standard.

The redox potential defines the energy level at which electrons are given out by the reduced form or accepted by the oxidized form of a given redox couple. The rule is that electrons fall spontaneously from a higher to a lower energy level, that is, move from the reduced form of the couple having the more negative, or less positive, redox potential to the oxidized form of the couple with the less negative, or more positive, redox potential. Such an electron transfer releases energy in an amount proportional to the difference between the two redox potentials and to the number of electrons transferred. For this energy to be equal to the bioenergetic unit of 14 kcal, or 59 kJ, per gram-molecule (see the section "Life's Jack of All Trades" in the preceding chapter), the potential difference must be at least 600 mV for single-electron transfers and 300 mV if the electrons travel in pairs.

In other words, to be able to power the assembly of one ATP molecule from ADP and inorganic phosphate, a single electron must fall down a potential difference of at least 600 mV. The minimum difference is 300 mV if, as is often the case, pairs of electrons are transferred. For the reverse reaction by which electrons are lifted to a higher energy level with the help of ATP, these potential differences represent the maximum energy boost electrons may gain, singly or in pairs, from the hydrolysis of one ATP molecule.

Bioenergetic Functions of Electron Transfers

Not all biological electron transfers play a bioenergetic role. Many take place in the course of metabolism and are governed by the simple thermodynamic rule that the donation of electrons must occur at a higher energy level than their acceptance. The difference in redox potential is often small, however, so that the transfers are readily reversed by an appropriate change in reactant concentrations.

When electron transfers serve to support life energetically by way of ATP, what is needed, as we have just seen, is an electron donor and an electron acceptor separated by a redox potential difference of at least $600/n$ mV (n being the number of electrons transferred), together with a catalytic system capable of mediating the transfer of electrons between donor and acceptor in a manner – the technical term is *coupling* – such that ATP is obligatorily assembled from ADP and inorganic phosphate. ATP, as we have seen, will do the rest, or almost. These requirements, represented schematically in Figure 5.1, are met differently in heterotrophic and autotrophic organisms.

As illustrated in Figure 5.2, heterotrophic organisms, which depend on other living organisms for their survival (*heteros* means "other" in Greek), derive their electron donors from food. Their final electron acceptor is most often molecular oxygen (O_2), which accepts four hydrogen atoms (electrons + protons) to form two molecules of water:

$$O_2 + 4\,(-) + 4\,H^+ \longrightarrow 2\,H_2O \qquad (1)$$

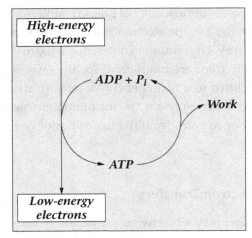

Figure 5.1. *The source of biological energy.* With very rare exceptions, the coupled assembly of ATP from ADP and inorganic phosphate is supported by the fall of electrons from a higher to a lower energy level. The splitting of ATP into ADP and inorganic phosphate, in turn, powers most forms of biological work (see Chapter 4). For the coupling to be thermodynamically feasible, the difference between the two energy levels (redox potentials) under physiological conditions must be equal to at least $600/n$ mV, n being the number of electrons transferred (1 or 2). Up to four such "electron falls," each coupled to the assembly of one ATP molecule, may take place in succession, for example when an electron pair travels from an aldehyde or α-keto acid to molecular oxygen (redox potential difference ≈ 1500 mV). This core mechanism operates in four different modes (see Figures 5.2 to 5.5), depending on the source and fate of the electrons.

Such is the case for all fungi, plants in the dark, and animals, ourselves included, and for many single-celled organisms, prokaryotic as well as eukaryotic. All these organisms belong to the group known as *aerobic* (living in air). They take maximum advantage of their environment in that the water/oxygen couple occupies the lowest energy level available to life.[1]

Some heterotrophic prokaryotes use an electron acceptor other than oxygen, for example, ferric iron (Fe^{3+}), which is reduced to ferrous iron, (Fe^{2+}), or elementary sulfur (S), which is reduced to hydrogen sulfide (H_2S). In the special case of anaerobic (without oxygen) fermentations, such as alcoholic or lactic fermentation, no outside electron acceptor is

---

[1] Strictly speaking, hydrogen peroxide (H_2O_2) is a stronger oxidant than oxygen, but its generation (by electron transfer to oxygen by type II oxidases) is essentially biological.

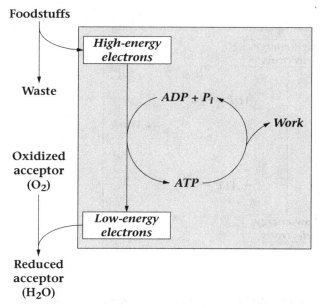

Foodstuffs

Waste

**Oxidized
acceptor
(O₂)**

**Reduced
acceptor
(H₂O)**

Figure 5.2. *Energetics of heterotrophy.* The core mechanism of Figure 5.1 is fed electrons at a high energy level by organic foodstuffs and returns the electrons at a low energy level to an outside acceptor, most often oxygen.

used at all. The acceptor arises from the donor by metabolism and is excreted in reduced form, for example, ethanol or lactate (see Figure 5.3).

In contrast to heterotrophs, autotrophic organisms are self-sufficient (*autos* means "self" in Greek), in the sense that they can survive in the absence of other forms of life. This prerogative is obtained at considerable energetic cost over and above the needs of heterotrophic life. Autotrophs not only have the same ATP needs as heterotrophs, similarly covering these needs by assembly reactions coupled to the fall of electrons from a higher to a lower level of energy, they must, in addition, manufacture all their constituents from simple substances of nonbiological origin, using carbon dioxide (CO_2) as a source of carbon, water (H_2O) as a source of hydrogen, nitrate (NO_3^-) or, sometimes, ammonia (NH_3), as a source of nitrogen, sulfate (SO_4^{2-}) or some other sulfur-containing mineral as a source of sulfur, and so on. Most of these building blocks are highly oxidized. Their conversion into organic molecules requires many reductions

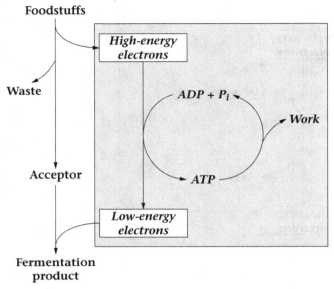

Figure 5.3. *Energetics of fermentation.* The core mechanism of Figure 5.1 is fed electrons at a high energy level by organic foodstuffs, as in all forms of heterotrophy (see Figure 5.2), but without the participation of an outside electron acceptor. The spent electrons are transferred to a metabolically generated acceptor, which is excreted in reduced form, making up the product of fermentation (for example, ethanol or lactic acid).

(biosynthetic reductions) and, therefore, a large supply of energy-rich electrons that are not needed for this function by heterotrophic organisms.[2] Such energy-rich electrons are not found in nature. Derived from low-energy sources, they must be actively lifted to the required energy level.

There are essentially two forms of autotrophy: chemotrophy and phototrophy. Chemotrophs, which are present only in the prokaryotic world, resemble heterotrophs in that they receive electrons at a high energy level from outside donors and return the electrons at a low energy level to an outside acceptor, most often oxygen (see Figure 5.4). The electrons, however, are provided not by organic foodstuffs but by simple inorganic

---

[2] Heterotrophic organisms do effect a small number of biosynthetic reductions. The conversion of ribose to deoxyribose and the formation of fatty acids from carbohydrates are conspicuous examples.

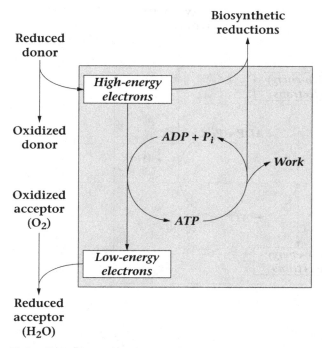

Figure 5.4. *Energetics of chemotrophy.* This mechanism functions like heterotrophy (see Figure 5.2), except that the electrons are provided by a mineral donor and that only part of them are transferred to an outside acceptor with the coupled assembly of ATP. The remaining electrons are used for biosynthetic reductions. In order to be able to perform this function, the electrons must be lifted to a higher energy level than is attained by the outside donors. This necessary energization is supported by ATP.

substances, for example, hydrogen sulfide. In addition, in order to satisfy the requirements of autotrophy, the organisms divert some of the electrons thus gained to cover biosynthetic reductions, first lifting the electrons to the necessary energy level with the help of ATP. Many forms of chemotrophy exist, depending on the nature of the electron donors and acceptors used. A special, particularly simple mode of chemotrophy is carried out by methane-producing prokaryotes. Known as methanogens, these organisms use molecular hydrogen (H_2) as electron donor and carbon dioxide (CO_2) as acceptor, to produce methane (CH_4) and water (H_2O), with the coupled assembly of ATP.

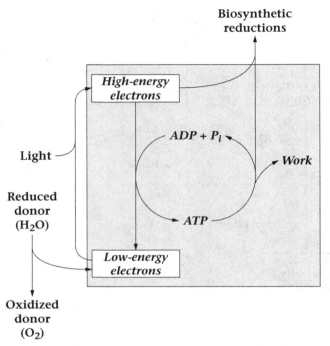

Figure 5.5. *Energetics of phototrophy.* The core mechanism of Figure 5.1 functions without either outside electron donor or outside electron acceptor; it is entirely powered by light energy, which lifts spent electrons back to a high energy level (cyclic photophosphorylation). Outside electrons (most often provided from water, with formation of molecular oxygen) are required for biosynthetic reductions, as in chemotrophy (see Figure 5.4). As in chemotrophy, these electrons require an additional energy boost from ATP to be utilizable.

Phototrophs function like chemotrophs except that they have the ability, dependent on the presence of the green pigment chlorophyll, to lift electrons to a high energy level with the help of light energy. They can thus power ATP assembly without the participation of outside electron donors or acceptors (see Figure 5.5), simply by recycling the same electrons endlessly with the help of light (cyclic photophosphorylation). Phototrophs do, however, need outside electrons to support biosynthetic reductions. In lower photosynthetic bacteria, the donor is some reduced mineral, as in chemotrophs, or sometimes even an organic compound, thus combining heterotrophy and phototrophy. In the higher photosynthetic organisms, including all green plants, the donor is water, with release of molecular oxygen [reaction (1) reversed]. After light activation, the electrons need an additional energy boost supported by ATP to serve for biosynthetic reductions, as in chemotrophs.

In summary, life derives the electrons it needs from at least one of three sources: organic foodstuffs, mineral substances, and water. In all organisms that are unable to use light energy (including phototrophs in the dark), the electrons used for ATP assembly are returned to the environment by transfer to an acceptor of lower energy level, most often oxygen, but in rare cases in combination with a metabolically generated acceptor (fermentations). Illuminated phototrophic organisms satisfy their energy needs without any outside electron donor or acceptor, relying for ATP assembly on the cycling of chlorophyll electrons between ground and activated level with the help of light energy. In addition, all autotrophs need outside electrons to support biosynthetic reductions. These electrons are never supplied at a high enough level of energy for immediate use. (In most cases, they are supplied at what may be termed bottom level, that of the water/oxygen couple.) The electrons are lifted to the energy level required for biosynthetic reductions either with the exclusive help of ATP splitting (chemotrophs) or with the help of light energy supplemented at the upper step by ATP splitting (phototrophs).

Catalysts

Consonantly with the crucial role of electron transfers in the economy of all living organisms, catalysts of such transfers make up an important fraction of any cell's enzymatic endowment. (The other major group is involved in group transfers; see Chapter 4.)

Electron-transfer machineries include not only enzymes, but also a number of cofactors, or coenzymes, that participate as electron carriers in the transfers mediated by the enzymes. Some of these coenzymes are universally distributed and may go back to very early days in the origin of life. Among them are substances known to biochemists under the acronyms NAD, NADP, FMN, and FAD, which consist of nitrogenous bases (N stands for nicotinamide, F for flavin) engaged in nucleotide-like combinations. Other electron carriers include quinones; iron–sulfur proteins; metal ions such as iron, manganese, and copper; and cytochromes. Cytochromes belong to the group of *hemoproteins*, which includes hemoglobin (the oxygen carrier of vertebrate blood) and a number of important enzymes. One such enzyme, *cytochrome oxidase* (appropriately called the "respiratory enzyme" by its discoverer, Otto Warburg), occupies a special place as the main oxygen consumer of the living world.

As will be seen in Chapter 11, electron carriers are often associated in complex chains embedded in the fabric of a membrane.

Coupling Mechanisms

Electron transfers are bioenergetically useful only if they are *coupled* to ATP assembly, that is, follow a pathway such that the electrons can pass from donor to acceptor only if ADP and inorganic phosphate join into ATP. Note that some of these coupled transfers are reversible and thus can, in the reverse direction, support the uphill transfer of electrons with the help of ATP splitting. This process, as we have seen, typically supports biosynthetic reductions.

Coupling mechanisms are highly complex and will be examined in separate chapters. Let it simply be mentioned that there exist two main such mechanisms. By far the most important one, from the quantitative point of view, involves an entity known as protonmotive force. The machinery involved in this process is obligatorily housed in the matrix of a membrane. We shall devote a special chapter to its description (see Chapter 11).

The second type of coupling mechanism is of minor importance quantitatively. Qualitatively, however, it is at least as important as the first, being universally distributed and involved in some of the most central and, probably, most ancient metabolic systems, including the glycolytic chain and the Krebs cycle. This mechanism involves thioesters, the topic of the next chapter.

The First Electron Transfers

When did electron transfers start playing a role in protometabolism, and what were the earliest mechanisms involved? These fundamental questions have rarely been addressed and, when addressed, have received conflicting hypothetical answers. These can be grouped into two categories, depending on the favored direction of the first electron transfer reactions.

We have seen in Chapter 1 that many of the building blocks of life arise spontaneously in outer space and could have provided the first materials for the origin of life on Earth, together, perhaps, with substances synthesized locally by the kind of processes Miller and others have tried to reproduce in the laboratory. This possibility, embodied in the so-called heterotrophic theory of the origin of life, implies that some of these products of cosmic and prebiotic Earth chemistry acted as the first electron donors, the way foodstuffs do in heterotrophs today, and that the electrons were caught by some mineral acceptors, though most probably not oxygen, which is generally taken to have been present in only trace amounts in the prebiotic atmosphere (see Chapter 16).

The postulated electron transfers would have required catalysts. But this does not raise insuperable difficulties. A number of metal ions, mineral iron–sulfur complexes (Cammack, 1983), or my hypothetical multimers could have done the job. A more challenging and totally unanswered question is how energy released by electron transfers first came to be usefully exploited by emerging life. We have seen in the preceding chapter that ATP and other NTPs presumably appeared early in protometabolism. They must have done so if they served as precursors in the first synthesis of RNA. But a connection between electron transfer and ATP is hardly evident. It seems unlikely that the kind of complex systems dependent on protonmotive force that operate today could have arisen, even in rudimentary form, in the early days of the development of life. A much more likely hypothesis is that the first couplings between electron transfer and ATP assembly occurred by way of thioesters. This topic will be covered in Chapter 6.

The heterotrophic theory of the origin of life is not unanimously accepted. Two investigators in particular, the German Günter Wächtershäuser (1998) and the American Harold Morowitz (1999), have vigorously advocated, though independently and on different premises, the notion that life started in an autotrophic manner. They both see this development as having taken place progressively around the Krebs cycle acting in reverse, that is, reductively. Leaving out details, a feature common to both theories – and to any other autotrophic theory of the origin of life – is that the first electron transfers must have taken place in the reductive direction, that is, from some energy-rich, environmental electron donor to some oxidized, biogenic acceptor, such as CO_2. Morowitz makes no special proposal concerning the nature of the donors, beyond identifying them tentatively as "simple reductants, such as formate and H_2S, or CH_3SH." Wächtershäuser, on the other hand, has considered two possibilities. In his first works, he identified the conversion of sulfide to disulfide as the electron-donating process:

$$2\ S^{2-} \longrightarrow S_2{}^{2-} + 2\ (-) \tag{2}$$

This reaction, which in itself is not very favorable thermodynamically, is taken to be driven by the presence of ferrous ions (Fe^{2+}), which trap the generated disulfide in the form of the highly insoluble ferrous disulfide, (FeS_2), the constituent of the mineral pyrite (fool's gold), which is assumed to catalyze the reaction. Experiments have shown that this process can indeed function as proposed.

The second process conjectured by Wächtershäuser, also confirmed experimentally, postulates carbon monoxide as electron donor, forming carbon dioxide. The interest of this reaction is that it can serve as condensing mechanism at the same time:

$$CO + X\text{-}OH + Y\text{-}H \longrightarrow CO_2 + X\text{-}Y + 2\ (-) + 2\ H^+ \tag{3}$$

The catalyst, in this case, consists of a mixture of ferrous and nickel sulfides, FeS/NiS. These reactions form the core of Wächtershäuser's "iron–sulfur world," a set of processes of increasing complexity assumed to have taken place in a volcanic setting, on the surface of pyrite crystals.

A number of workers have envisaged the possibility that light may, one way or the other, have served to power an autotrophic beginning of life. Besides including few details concerning the mechanisms involved in the harnessing of light energy and in the development of the first metabolic reactions, such proposals suffer from the basic defect that they rely, by definition, on an intermittent supply of energy. How the processes could have survived regular periods of darkness has not been addressed.

Future work will perhaps help clarify these fascinating problems. At present, the main weakness of the autotrophic theory seems to me to be that it fails to take into consideration the possible contributions of cosmic (and terrestrial prebiotic) chemistry. As argued in Chapter 1, the fact that so many of the basic building blocks of life arise spontaneously can hardly be attributed to a meaningless coincidence. In particular, it does not seem logically justified to ignore the possible cosmic (and prebiotic) origin of the first amino acids and to attribute their formation instead to amination reactions grafted on a primordial Krebs cycle.

The fact remains that autotrophy must have been developed at the latest when the supply of organic chemicals became exhausted. This means that, by that time, some system must have been developed, not only for energy generation by downhill electron transfer, but also for powering the uphill electron transfers required for the reduction of carbon dioxide and the other mineral precursors of biological molecules. As I have advocated for more than fifteen years, a possible answer to both this and the former question lies in one word: thioesters. The point is sufficiently important to deserve a separate chapter.

6 Thioesters

Thioesters are the products of the dehydrating condensation of a carboxylic acid (R′–COOH) and a thiol (R′–SH):

$$\underset{\text{R-C-OH}}{\overset{\overset{\textstyle O}{\|}}{}} + \text{R′-SH} \longrightarrow \underset{\text{R-C-S-R′}}{\overset{\overset{\textstyle O}{\|}}{}} + H_2O \qquad (1)$$

The CO–S bond is called the thioester bond. Its energy value is equal to that of the pyrophosphate bond. This means (see "Life's Jack of All Trades" in Chapter 4) that reaction (1), like ATP assembly from ADP and P_i, requires about 14 kcal, or 59 kJ, per gram-molecule to take place in living cells. Its reversal, thioester hydrolysis, releases the same amount of energy.

Thioesters stand out by their metabolic position as reversible bridges between electron transfers (Chapter 5) and group transfers (Chapter 4). They are truly unique in this respect, offering one of the most noteworthy singularities displayed by life. Their essential properties will be briefly summarized in this chapter. Readers wishing for more details are referred to my previous publications on the topic (de Duve, 1991, 1998, 2001).

Thioesters and Electron Transfers

The thiol group plays a central role in the biological oxidation of carbonyl (=C=O) to carboxyl (–COOH) groups. Two main types of such reactions are known. In one type, an aldehyde is oxidized to the corresponding acid:

$$R\text{-}\overset{\overset{\displaystyle O}{\|}}{C}\text{-}H + H_2O \longrightarrow R\text{-}\overset{\overset{\displaystyle O}{\|}}{C}\text{-}OH + 2\ (-) + 2\ H^+ \tag{2}$$

The other type of reaction is the oxidative decarboxylation of an α-keto acid:

$$R\text{-}\overset{\overset{\displaystyle O}{\|}}{C}\text{-}\overset{\overset{\displaystyle O}{\|}}{C}\text{-}OH + H_2O \longrightarrow R\text{-}\overset{\overset{\displaystyle O}{\|}}{C}\text{-}OH + CO_2 + 2\ (-) + 2\ H^+ \tag{3}$$

In the biological versions of these reactions, the electrons released are donated to NAD, which accepts electrons roughly 300 mV above the potential (that is, below the energy level) at which they are donated by the carbonyl donors. It follows (see "Energetics of Electron Transfer" in Chapter 5) that, with NAD as electron acceptor, reactions (2) and (3) release an amount of energy on the order of the bioenergetic unit of 14 kcal, or 59 kJ, per gram-molecule [somewhat more for reaction (3) than for reaction (2)].

A key feature of these electron transfers when they take place in living cells is that a thiol belonging to the cells' catalytic machinery (R'–SH) substitutes for water, so that a thioester is formed instead of the free acid:

$$R\text{-}\overset{\overset{\displaystyle O}{\|}}{C}\text{-}H + R'\text{-}SH \longrightarrow R\text{-}\overset{\overset{\displaystyle O}{\|}}{C}\text{-}S\text{-}R' + 2\ (-) + 2\ H^+ \tag{4}$$

$$R\text{-}\overset{\overset{\displaystyle O}{\|}}{C}\text{-}\overset{\overset{\displaystyle O}{\|}}{C}\text{-}OH + R'\text{-}SH \longrightarrow R\text{-}\overset{\overset{\displaystyle O}{\|}}{C}\text{-}S\text{-}R' + CO_2 + 2\ (-) + 2\ H^+ \tag{5}$$

Some simple bookkeeping will show that reaction (4) equals the sum of reactions (2) and (1), and that reaction (5) equals the sum of reactions (3) and (1). In other words, the energy released by the electron transfers [reactions (2) and (3), with NAD as electron acceptor] is used to create a thioester bond between the acid product of the reaction and the participant thiol [reaction (1)].

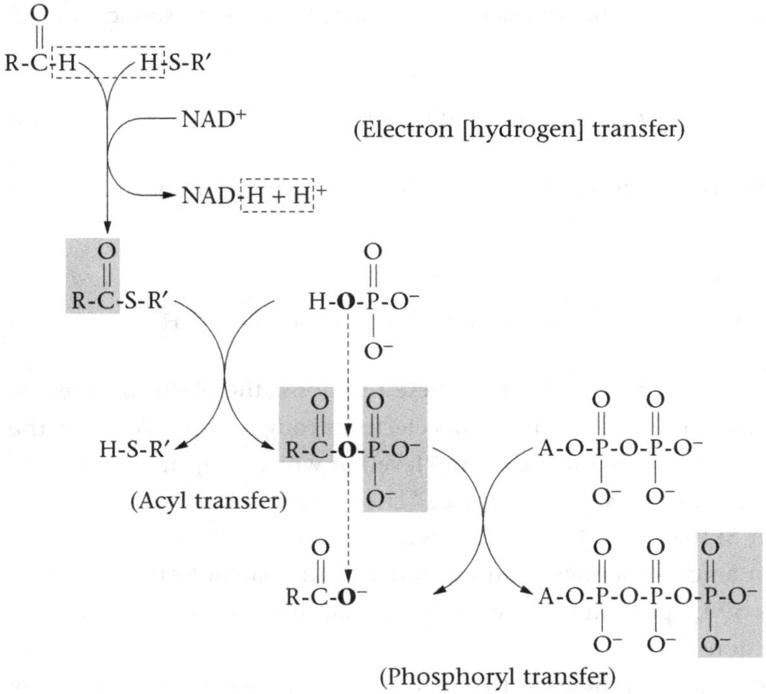

Figure 6.1. *Thioester-mediated coupling between the oxidation of an aldehyde and ATP assembly.* In the first step, the aldehyde and the thiol *jointly* donate a pair of hydrogen atoms (two electrons and two hydrogen ions) to the oxidized form of the coenzyme NAD, with the resulting formation of a thioester and the reduced form of NAD. This reaction corresponds to reaction (4) in the text. Its crucial role is to create an energy-rich bond (the thioester bond) with the help of an electron transfer; it conserves in this bond the energy released by the electron transfer, which occurs across a redox potential difference on the order of 300 mV. The other two reactions show the assembly of ATP from ADP and inorganic phosphate at the expense of the splitting of the thioester bond, by sequential group transfer. They correspond to the reverse of the scheme shown in Figure 4.1, with a carboxylic acid as X–OH and an acyl phosphate as double-headed intermediate. Note the pathway of the oxygen atom from inorganic phosphate to the carboxyl group of the acid.

The importance of such processes can hardly be overestimated. They allow the direct formation of a chemical bond with the help of energy released by electron transfer. The significance of this singularity is enormously enhanced by the fact that thioesters can support the reversible

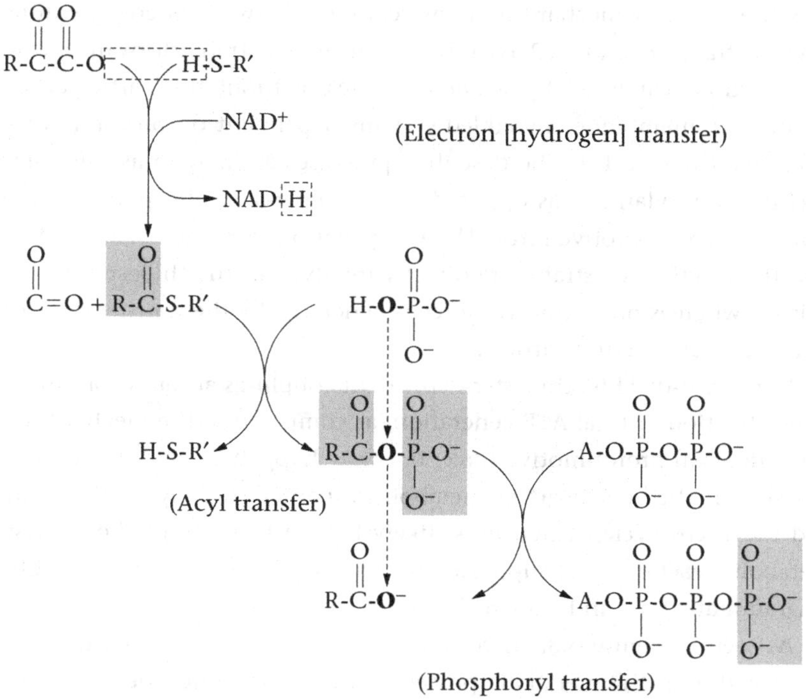

Figure 6.2. *Thioester-mediated coupling between the oxidative decarboxylation of an α-keto acid and ATP assembly.* The process takes place as in the case of an aldehyde (Figure 6.1), except that the first step corresponds to reaction (5). The other steps are identical.

assembly of ATP by sequential group transfer (using the reverse of the process illustrated in Figure 4.1, with a carboxylic acid as X–OH reactant). Written in shorthand,

$$R\text{-CO-S-}R' + P_i \longrightarrow R\text{-CO-P} + R'\text{-SH} \qquad (6)$$

$$R\text{-CO-P} + ADP \longrightarrow R\text{-COOH} + ATP \qquad (7)$$

$$\overline{R\text{-CO-S-}R' + P_i + ADP \longrightarrow R\text{-COOH} + R'\text{-SH} + ATP} \qquad (8)$$

Putting together reaction (4) or (5) with reactions (6) and (7), we have the diagrams of Figures 6.1 and 6.2, which show explicitly how the thioester mechanism allows coupling between energy-yielding electron transfers and ATP assembly by means of straightforward *chemical* reactions catalyzed by *soluble* enzymes, without the participation of complex membrane-embedded systems dependent on protonmotive force (see Chapter 11). The described processes are known as *substrate-level* phosphorylations, as opposed to the *carrier-level* phosphorylations involving protonmotive force. These appellations are understandable in that the reaction substrates are directly involved in the thioester mechanisms, whereas protonmotive force is generated by the transfer of electrons between electron carriers.

In present-day life, thioester-dependent couplings account for only a minor fraction of total ATP generation, as compared to the mechanisms dependent on protonmotive force; but they happen to be directly associated with the very heart of metabolism, namely the glycolytic chain and the Krebs cycle, which most likely belong to some of the earliest metabolic reactions that supported nascent life. This is shown in highly diagrammatic form in Figure 6.3.

As seen in Figure 6.3, three thioester-mediated systems are directly associated with the metabolic processes. In the first one, the oxidation of phosphoglyceraldehyde to phosphoglycerate, with NAD as electron acceptor, is linked with the assembly of ATP. It takes place by the mechanism of Figure 6.1, the thiol R'-SH being provided by a cysteine residue of the enzyme protein. Note that this reaction represents the sole energetic benefit cells derive from the anaerobic fermentation of glucose. In this process, pyruvate serves as electron acceptor in the reoxidation of NADH, to form either lactate or ethanol and CO_2, which are excreted as the final fermentation products.

The two other systems, associated with the oxidative decarboxylation of pyruvate on one hand, and of α-ketoglutarate on the other, follow the mechanism outlined in Figure 6.2, with, as thiol, a substance known as coenzyme A. We shall see later the structure of this molecule, which will be represented provisionally simply as CoA–SH. Not shown in the figure is the intervention of an additional substance, called lipoic or thioctic acid, an eight-carbon carboxylic acid bearing two thiol groups, which can suffer

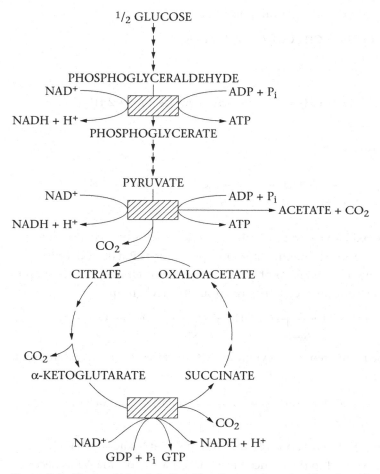

Figure 6.3. *The heart of metabolism*. The diagram shows in abbreviated form the two central metabolic processes that occur in the vast majority of living organisms: above from glucose to pyruvate, the glycolytic chain; below, the Krebs cycle. The three shaded boxes represent three thioester-mediated coupling mechanisms. The first one functions according to the scheme of Figure 6.1; the other two, according to the scheme of Figure 6.2 (G replacing A in the third one). Note that two outlet systems are depicted for the middle thioester system. Either acetate and CO_2 are formed, with coupled assembly of ATP, or the acetyl group is combined with oxaloacetate to form citrate, initiating the Krebs cycle. Additional details are given in the text.

oxidation to the corresponding disulfide (S–S):

$$CH_2\text{-}CH_2\text{-}\underset{\underset{SH}{|}}{C}H\text{-}CH_2\text{-}CH_2\text{-}CH_2\text{-}CH_2\text{-}COOH$$

where SH is also on the first carbon.

$$\downarrow$$

$$CH_2\text{-}CH_2\text{-}CH\text{-}CH_2\text{-}CH_2\text{-}CH_2\text{-}CH_2\text{-}COOH + 2\,(-) + 2\,H^+$$

$$\underset{S\text{———}S}{|\quad\quad|}$$

(9)

In simpler notation,

$$\underset{SH\ SH}{\boxed{}}\text{-}COOH \longrightarrow \underset{S\text{—}S}{\boxed{}}\text{-}COOH\ + 2\,(-) + 2\,H^+$$

In the oxidative decarboxylation of α-keto acids (which also requires another coenzyme, thiamine pyrophosphate, or TPP, a derivative of vitamin B_1), the oxidized form of lipoic acid serves as electron acceptor and, at the same time, as acceptor of the resulting acyl group:

$$R\text{-}CO\text{-}COOH + \underset{S\text{—}S}{\boxed{}}\text{-}COOH \longrightarrow \underset{SH\ S\text{-}CO\text{-}R}{\boxed{}}\text{-}COOH + CO_2 \qquad (10)$$

In a subsequent step, the acyl group is transferred to the thiol group of coenzyme A:

$$\underset{SH\ S\text{-}CO\text{-}R}{\boxed{}}\text{-}COOH + CoA\text{-}SH \longrightarrow \underset{SH\ SH}{\boxed{}}\text{-}COOH + CoA\text{-}S\text{-}CO\text{-}R \qquad (11)$$

Sum reactions (10) and (11), and you find that lipoic acid has changed from the oxidized to the reduced form and that coenzyme A has become the carrier of the acyl product of the reaction. Note that the reduced lipoic acid is reoxidized, ready for another round, by transfer of the electrons to NAD:

$$\underset{SH\ SH}{\boxed{}}\text{-}COOH + NAD^+ \longrightarrow \underset{S\text{—}S}{\boxed{}}\text{-}COOH + NADH + H^+ \qquad (12)$$

NAD thus ends up accepting the electrons released, as shown in the diagram of Figure 6.3.

The unique role of lipoic acid deserves to be emphasized. It truly embodies the quintessence of thioester biochemistry in that it joins, in a single molecule of simple structure, the two main processes of biological energy transfer: electron transfer and group transfer. As far as I know, no other comparable substance exists.

The thioesters formed by the oxidative decarboxylation of pyruvate (acetyl CoA) or α-ketoglutarate (succinyl CoA) may support the assembly of ATP (or GTP, when α-ketoglutarate is the substrate), as shown in Figure 6.3. But they can also become involved in other acyl transfer reactions. As we shall see presently, coenzyme A is a major group carrier involved in numerous biosynthetic processes. In particular, acetyl CoA, the product of the oxidative decarboxylation of pyruvate, supports ATP assembly only in certain microbes. Its principal metabolic function (also indicated in Figure 6.3) is to react with oxaloacetate to form citrate, thereby connecting glycolysis with the Krebs cycle. Thus, whenever glycolysis leads into the Krebs cycle, only two of the three thioester mechanisms serve in the assembly of ATP (or GTP). The third one (middle in Figure 6.3) provides the energy needed for the formation of citrate.

Note that this is only little more than one-twentieth of the benefit aerobic organisms derive from these systems today. When glucose is oxidized with oxygen as final electron acceptor, altogether 34 ATP molecules are assembled per molecule of glucose oxidized, as opposed to 2 by thioester mechanisms.[1] But the other 32 are assembled with the help of protonmotive force. If we take as likely that protonmotive force, with all the complex membrane-embedded systems it involves, is a relatively late development, then the mechanisms mediated by thioesters are the only ones among those operating today that could have supported the early couplings between electron transfer and ATP assembly.

These mechanisms also play a crucial role in biosynthetic reductions. Looking at the diagram of Figure 6.3, one readily sees that changing the direction of the arrows allows the conversion of three molecules of CO_2 to half a molecule of glucose. Defenders of an autotrophic origin of life attach considerable significance to the fact that certain chemotrophic bacteria use the reverse Krebs cycle for biosynthetic reductions. But such cases

[1] In reality, the gross figures are 36 in the first case, and 4 in the second. But deduction must be made of the two ATP molecules invested in the phosphorylation reactions that spark glycolysis.

are exceptional. Of much more general importance is reverse glycolysis, which plays a key role in most forms of autotrophy. It was mentioned in the preceding chapter that biosynthetic reductions include one step that requires a boost from ATP, even with activated chlorophyll as electron donor. This step is the reduction of phosphoglycerate to phosphoglyceraldehyde in reverse glycolysis (see Figure 6.3). As just seen, this step involves a thioester-mediated mechanism. Thus, there would be no autotrophy without thioesters.

Thioesters and Group Transfers

As was mentioned in Chapter 4, biosynthetic assemblies invariably require activation of one of the reactants to overcome the energy barrier that opposes the reaction. In a large number of processes involving carboxylic acids, these molecules are activated in the form of thioesters. Among the substances that depend on thioesters for their synthesis are fatty acids, lipid esters, citrate (in the Krebs cycle), malate (in the glyoxylate cycle), porphyrins, and terpenes (including sterols), as well as many nonprotein peptides and polyketides (see the last section of Chapter 3).

The main carrier involved in these processes is a molecule known as pantetheine phosphate, acting either in protein-bound form or as part of coenzyme A, in which it is linked with 3'-phospho-AMP. The structure of pantetheine is given below:

$$\text{HO-CH}_2\text{-CH}\underset{\underset{\text{CH}_3}{|}}{\text{—}}\text{CH}\underset{\underset{\text{CH}_3}{|}}{\text{-}}\overset{\overset{\text{O}}{\|}}{\text{C}}\text{-NH-CH}_2\text{-CH}_2\text{-}\overset{\overset{\text{O}}{\|}}{\text{C}}\text{-NH-CH}_2\text{-CH}_2\text{-SH}$$

It is made by the joining, by two CO–NH bonds, of three simple molecules: hydroxydimethylbutyric acid, β-alanine, an amino acid (not involved in protein synthesis), and cysteamine, the product of decarboxylation of another amino acid, cysteine. In pantetheine phosphate, the phosphate group is attached to the terminal OH group of the molecule. In coenzyme A, this phosphate is linked to the phosphate of AMP, which carries an additional phosphate group in position 3' of the ribose.

When not resulting from electron transfers, the attachment of acids to the thiol group of their carrier is supported by ATP, by way of sequential group transfer. In a few cases, the terminal pyrophosphate bond of ATP provides the energy [reaction (8) reversed]. More frequently, the energy comes, by nucleotidyl transfer, from the splitting of the inner pyrophosphate bond of ATP (see Figure 4.3), which, as we have seen, is often used to drive particularly important and energetically costly biosynthetic processes:

$$ATP + R\text{-}COOH \longrightarrow PP_i + R\text{-}CO\text{-}AMP \qquad (13)$$

$$\underline{R\text{-}CO\text{-}AMP + CoA\text{-}SH \longrightarrow CoA\text{-}S\text{-}CO\text{-}R + AMP} \qquad (14)$$

$$ATP + R\text{-}COOH + CoA\text{-}SH \longrightarrow CoA\text{-}S\text{-}CO\text{-}R + AMP + PP_i \qquad (15)$$

Why Sulfur?

Or, rather, why hydrogen sulfide? Sulfate, the main form of sulfur on the present-day Earth, is of limited use to life; it serves as electron acceptor for some bacteria and provides certain macromolecules, mostly polysaccharides, with acidic side groups. Thiols are derivatives of hydrogen sulfide (H_2S), a gas with the characteristic stench of rotten eggs, familiar to generations of chemists for its use in the selective precipitation of heavy metals, and to hosts of tourists as a distinctive emanation of volcanic soils. It probably also pervaded the cradle of life.

There is much to be said for an early intervention of thiols and thioesters in the development of life. Some of the key molecules involved have relatively simple structures, possibly within the range of a primitive prebiotic chemistry. In fact, the three constituents of pantetheine have been obtained under plausible prebiotic conditions in Miller's laboratory and found to assemble correctly merely upon concentration by evaporation (Keefe et al., 1995). Furthermore, the metabolic processes in which thiol derivatives are involved are catalyzed by soluble enzymes and take place by straightforward chemical reactions, without all the complexities attending the generation and exploitation of proton potentials. In spite of those (relative) simplicities, thioesters participate in processes that are truly crucial, being concerned with both group and electron transfers and with the coupling between the two. Also impressive is the fact that some of these processes belong to the most basic and, presumably, most ancient metabolic pathways of universal significance, such as glycolysis and the Krebs cycle. Finally, it is worth remembering that many bacterial peptides are made at the expense of the thioesters of the relevant amino acids; so, too, could my hypothetical multimers have been made.

All these facts point to a volcanic setting for the origin of life. The same setting has been suggested earlier in connection with the availability of inorganic pyrophosphate and polyphosphates. I shall come back to this possibility in a later chapter. In the meantime, it is interesting to speculate on the manner in which the two basic chemistries of life, phosphate chemistry and sulfur chemistry, could have come to be intermixed. Based on what is known today, the simplest mechanism that comes to mind consists of the reversible interaction of a thioester with inorganic phosphate leading to the assembly of inorganic pyrophosphate:

$$R\text{-CO-S-}R' + P_i \longrightarrow R\text{-CO-P} + R'\text{-SH} \tag{16}$$

$$R\text{-CO-P} + P_i \longrightarrow R\text{-COOH} + PP_i \tag{17}$$

$$R\text{-CO-S-}R' + 2\,P_i \longrightarrow R\text{-COOH} + R'\text{-SH} + PP_i \tag{18}$$

As shown, a thioester powers the assembly of pyrophosphate. Note that ATP will be assembled instead of inorganic pyrophosphate if P_i is replaced by ADP in reaction (17), or if P_i is replaced by AMP in reaction (16) and by PP_i in reaction (17), thus providing a possible bridge between pyrophosphate and ATP as central energy currency. On the other hand, the reactions shown all being reversible, changing their direction will have pyrophosphate (or ATP) supporting the synthesis of a thioester.

A Word about Iron

Mentioning sulfur brings up the topic of iron. This metal is associated with sulfide derivatives in iron–sulfur proteins, which, in view of their simple structure, could well number among the most primitive electron-transfer catalysts, possibly preceded by mineral iron–sulfur complexes (Cammack, 1983). Perhaps iron was ushered into protometabolism by sulfur compounds, eventually to become the central element of biological electron transfer. Today, this metal forms, in combination with a porphyrin, the characteristic heme group of cytochromes and other hemoproteins, in which the oscillation of the central iron atom between the ferrous and the ferric form determines electron donation and acceptance.

As mentioned in Chapter 5, iron and sulfur are intimately implicated in the model of the origin of life proposed by Wächtershäuser (1998).

7 RNA

It is generally accepted that RNA came before DNA in the origin of life. What we know of present-day life gives strong support to that contention. In all living organisms, DNA's sole function is to serve as the replicable repository of genetic instructions. When it comes to the execution of these instructions, RNA is the indispensable intermediary, occasionally as catalyst (ribozyme), most often as messenger (mRNA) for the synthesis of a protein that operates as the actual agent. DNA, in other words, can do nothing without RNA. In fact, DNA cannot even replicate without the prior formation of an RNA primer.[1] RNA, on the other hand, while being the obligatory instrument of all forms of genetic expression, does, in addition, share with DNA the property of replicability. This property is not manifested in normal cells but is realized in cells infected by certain viruses, such as the polio virus. Thus, we arrive at the almost inevitable conclusion that RNA was first on the scene, acting in the storage and replication of early genetic information as well as in its expression, until DNA appeared and took over the first functions.

There are solid reasons to believe that RNA preceded proteins as well. This claim, again, rests on our knowledge of present-day life, which universally depends on RNA molecules – tRNAs, mRNAs, and rRNAs – for the

---

[1] This rule is not absolute. Archaea replicate their DNA without an RNA primer. I am indebted to Günter Wächtershäuser for calling my attention to this fact.

synthesis of proteins. To be sure, life also uses proteins for the synthesis of RNA – and for that of proteins – presenting us with a typical chicken-or-egg problem. However, the role of RNA in protein synthesis is so important and specific, involving not only catalysis but also information, that the hypothesis of proteins arising first by some other mechanism and serving to make RNA, which would then subsequently take over their synthesis, seems very unlikely. But watch out! When speaking of proteins, we must recall that this term applies strictly to polypeptides made from twenty specific kinds of amino acids. With one exception (glycine), these are all chiral and of the L variety. That other peptides may have formed prebiotically by mechanisms not involving RNA is quite possible – I am inclined to say probable. My hypothetical multimers are a case in point (see Chapter 3).

The vast majority of scientists who have thought about the problem agree on the above points. But the mechanisms whereby the first RNA molecules may have emerged remain unknown and a matter for debate. Before addressing the question of origin, let us look first at what arose from it.

RNA Today

RNA, or ribonucleic acid, molecules are long strings, known as polynucleotides, assembled from four different kinds of units called nucleotides, or nucleoside monophosphates. These building blocks are AMP (adenosine monophosphate), of which much has been said already (see Chapter 4), and, briefly encountered before also, GMP, CMP, and UMP. These initials stand for guanosine monophosphate, cytidine monophosphate, and uridine monophosphate, respectively. Guanosine (G), cytidine (C), and uridine (U) are, like adenosine (A), combinations of nitrogenous bases – called guanine, cytosine, and uracil – with D-ribose. Such combinations are called nucleosides, and their monophosphates mononucleotides, nucleotides for short.[2] The structures of the four NMPs are illustrated in Figure 7.1.

---

[2] In the chemical nomenclature, the initials A, G, C, and U (and T) stand strictly for the nucleosides. They are often used in molecular biology to designate just the bases, which are the bearers of the biological information. This fuzziness rarely leads to confusion.

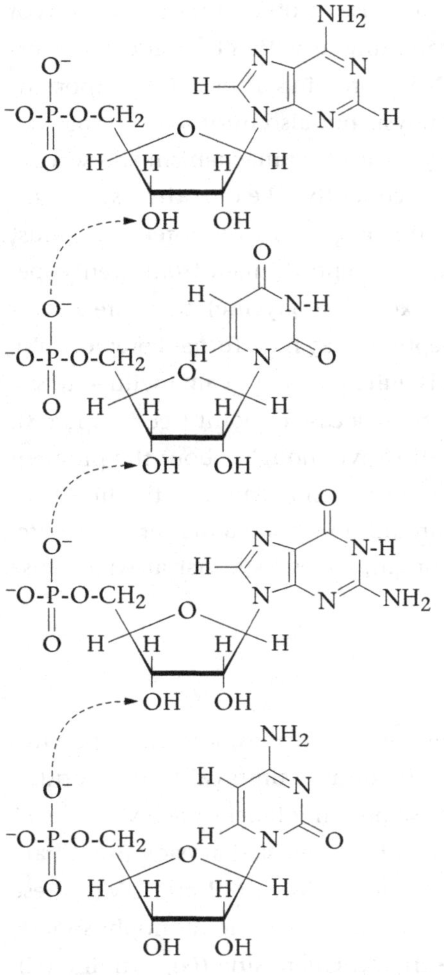

Figure 7.1. *Chemical structures of the four nucleotides used in RNA synthesis.* Curved broken arrows show where, in the formation of RNA molecules, the terminal phosphate group of one nucleotide joins with the 3′-hydroxyl group in the ribose molecule of an adjacent nucleotide (with release of OH^-).

In the biological synthesis of RNA (see Figure 7.2), the nucleotides are offered to the catalytic assembly machinery in the activated form of their pyrophosphate derivatives, the nucleoside triphosphates ATP, GTP, CTP, and UTP. In this process, the monophosphate parts of the triphosphates are attached one by one by their phosphate end to the ribose of

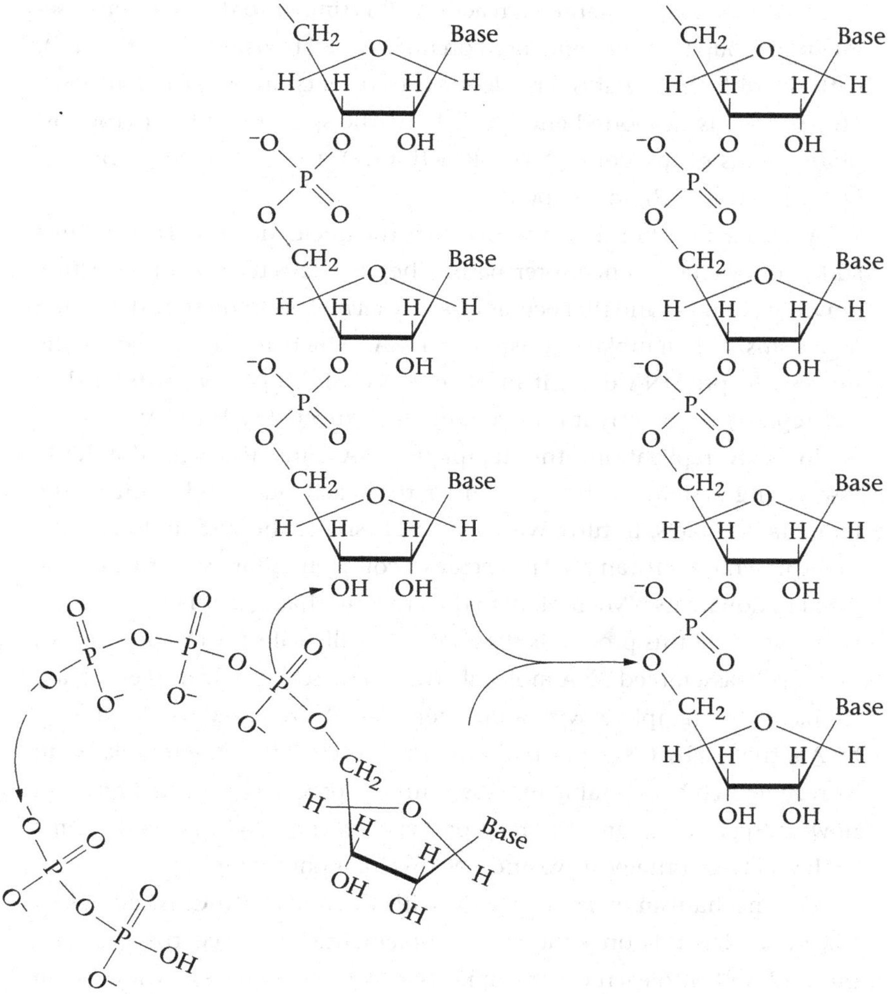

Figure 7.2. *Mechanism of chain elongation in RNA assembly.* The scheme shows how an NTP transfers an NMP unit to the end of a growing RNA chain, with liberation of inorganic pyrophosphate.

the terminal nucleotide of the growing polynucleotide chain, while the two terminal phosphates of the triphosphates are released as inorganic pyrophosphate, which is usually split to two phosphate molecules:

$$RNA + NTP \longrightarrow RNA \text{-} NMP + PP_i \tag{1}$$

$$PP_i + H_2O \longrightarrow 2\,P_i \tag{2}$$

As mentioned in Chapter 4, reaction (1) is similar to the first step of sequential group transfer dependent on nucleotidyl transfer (see Figure 4.3), with the difference that the nucleotidyl transfer completes the synthesis. This process is supported energetically by the splitting of the pyrophosphate bonds of the donor NTPs. Reaction (2) has already been encountered [reaction (13), in Chapter 4].

A crucial aspect of this process is that the nucleotide units are not just added on a chance-encounter basis. They are selected by a preexisting polynucleotide strand that acts as model, or *template*. In present-day living organisms, this template consists of DNA, which is transcribed in the process. In pre-DNA days, it most likely consisted of RNA, which thus was replicated, the way it is in certain virus-infected cells today.

In RNA replication, the template RNA runs through the RNA-assembling system in such a manner that each nucleotide unit of the template interacts, in turn, with a strategic site of the system, to create a niche in which only an NTP molecule of complementary structure can be fitted to donate its NMP moiety to the growing chain (see Figure 7.3). The rule governing this process is that A reciprocally calls for U, and G for C. The newly assembled RNA molecule thus has a structure complementary to that of the template, with all the template A's replaced by U's, the U's by A's, the G's by C's, and the C's by G's. It is clear that the template can be regenerated by the same process, with the product of the first reaction now acting as template. The relation between the two forms is the same as that between the positive and negative of a photograph.

The mechanism whereby A joins with U, and G with C, is called *base pairing*. It depends on structural complementarities such that the two molecules fit into each other, somewhat like puzzle pieces – they are, in fact, flat – and stick together by weak electrochemical attractions called hydrogen bonds (see Figure 7.4).

In addition to serving in RNA replication (by viruses), base pairing also governs the manner in which short, complementary RNA stretches join with each other. When these stretches belong to the same molecule, the outcome is folding of the chain into tridimensional shapes that may be highly complex. The most important instance of the joining of two distinct RNA molecules occurs in protein synthesis, with the interactions

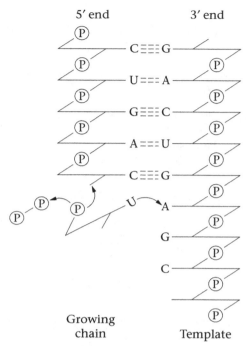

Figure 7.3. *Schematic representation of template-directed RNA assembly.* To the elongation step shown in Figure 7.2 has been added the intervention of a template that commands, by base pairing, the choice of the NMP unit to be hooked on to the growing chain.

between mRNA codons and tRNA anticodons whereby translation of the genetic information into the corresponding proteins is accomplished (see Chapter 8).

As we shall see in Chapter 9, base pairing also rules the structure of DNA – the famous "double helix," in which the phenomenon was first discovered – as well as its replication, the main difference being that U is replaced by T (thymine), which has the same configuration in the part of the molecule that pairs with A (see Figure 7.4). The same relationships function, with U and T appropriately standing for each other, when RNA is assembled on a DNA template (transcription) and when DNA is assembled on an RNA template (reverse transcription).

In summary, base pairing determines the shapes of all the DNAs and RNAs in the living world and, furthermore, governs all the basic biological

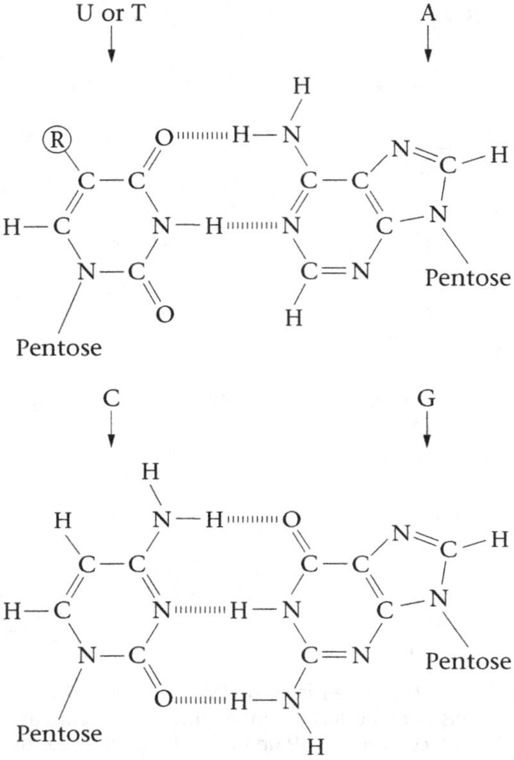

Figure 7.4. *Base pairing, the universal mechanism of biological information transfer.* On top is shown the pairing between adenine and uracil (R = H) or thymine (R = CH$_3$). The pairing between guanine and cytosine is shown below. Each base pair forms a flat plate, roughly elliptical in shape, with axes of 1.1 and 0.6 nm, and thickness 0.34 nm. Shaded bars are hydrogen bonds. Note that uracil, thymine, and cytosine are derivatives of the single-ringed pyrimidine nucleus. Adenine and guanine are derived from the purine nucleus, in which a pyrimidine nucleus is fused with an imidazole nucleus. Pentose is ribose in RNA derivatives, deoxyribose in DNA derivatives.

information transfers involving those molecules, whether in replication, transcription, or translation. Base pairing accomplishes those fundamental functions following two rules that are at the same time universal and almost ludicrously simple – to wit: A = U or T; G = C – the rules themselves being based on relatively weak electrostatic attractions between two chemically complementary structures. Base pairing truly represents one of the most astonishing singularities offered by present-day life. Most likely, this phenomenon appeared when life first "invented" RNA.

The Origin of RNA

No problem related to the origin of life has stimulated more research, fed more discussions, or evoked more passion than the first appearance of RNA, which is rightly viewed as a watershed, most likely the key event through which information first entered emerging life. In spite of all those efforts, only limited success has been obtained in attempts to reproduce RNA synthesis under prebiotic conditions. Short RNA-like chains have been assembled by means of mineral catalysts, mostly clays, with artificially activated nucleotides as precursors and with a few selected templates. The natural precursors have proved less effective, however, and the synthesis of the precursors themselves under plausible conditions has so far defied the ingenuity of investigators. A growing number of researchers are now turning from RNA, which they see as too complex to arise *de novo* under prebiotic conditions, to simpler molecules that could allegedly have preceded RNA as the first repository of genetic information (Joyce, 2002). A further examination of this approach would fall outside the scope of this book. Let it simply be stated that present-day life – our main guide, in line with the congruence notion – holds no evidence of the participation of simpler "pre-RNAs."

RNA synthesis has been a particularly favorable case for invoking the "fantastic luck" hypothesis (mechanism 6), the contention being that all that was needed was the appearance, by what could have been a highly improbable chance event, of just a few RNA molecules, which would then have been perpetuated by "self-replication." This could be a paradigmatic case of a transition, such as has often been evoked by theoreticians, from chaos to order by a chance fluctuation.

The notion of self-replication is misleading, however. In nature, RNA – or DNA, for that matter – does not truly self-replicate. It only serves as a passive template for an elaborate chemical catalytic machinery that puts together a complementary molecule from precursors that are themselves the products of complex chemical processes. What the hypothesis thus implies is that the extraordinary stroke of luck did not just produce any kind of RNA molecule but produced an RNA molecule capable of catalyzing its own replication. This amounts to the combination of two highly improbable lucky shots, putting a severe strain on plausibility. Not only is a chemically unlikely molecule required, but this molecule must have had a structure rendering it capable of catalyzing the template-directed assembly of its own kind. Implausible? Yes, but perhaps not impossible, as indicated by the successful design, by powerful *in vitro* selection techniques, of a ribozyme actually capable of copying a template with close to 99% accuracy, using the natural NTPs as precursors (Johnston et al., 2001). This is a remarkable performance, but still a long way short of the target. The maximal length of the synthesized RNA was only 14 nucleotides, which is not much compared to the thousands of nucleotides that can be correctly assembled today. Furthermore, as the authors readily admit, all that has been established is that an RNA-replicating RNA (though not yet self-replicating) can be created artificially by means of highly contrived, goal-directed manipulations. Nothing proves that such a molecule ever arose spontaneously by natural processes.[3] In addition, solving replication would not elucidate the mechanism of synthesis itself. There still would be left the problem of the prebiotic RNA precursors: their nature and their mode of formation.

Duly considered, the possibility that RNA may be the self-perpetuating product of a lucky accident can be rejected as putting implausible demands on chance, while at the same time stifling research. Unless one wishes to invoke the similarly stifling intelligent-design hypothesis, one must fall back on the view that prebiotic conditions were such as to favor the synthesis and replication of RNA molecules by straightforward chemical processes. At first sight, however, this hypothesis also seems to strain credibility.

But this may be only because we have been holding the wrong end of the stick. The mistake is to focus on RNA. Viewed in retrospect, the appearance of this substance does indeed strike us as a crucial watershed. But an observer present at the time would most likely not have seen it that way, accepting the event simply as an unsurprising manifestation of the existing chemistry. Chemistry has no prescience. It could not have

---

[3] Note that the autocatalytic participation of RNA molecules in their own lengthening, which could have taken place in the earliest phase of RNA development (see the last section of this chapter), remains a valid possibility.

produced NTPs, or whatever molecules served as the first building blocks of RNA, "for the purpose" of using them to synthesize RNA, itself "destined" to be the first bearer of information in nascent life. Unless we accept intelligent design, it is clear that the RNA precursors must have arisen spontaneously as a result of existing conditions. The "secret" of RNA is hidden in protometabolism. This, at least, seems to me a much more probable and fruitful hypothesis than the opposite assumption that RNA arose first by some unlikely combination of circumstances and gave rise later to what are now its precursors.

The Protometabolic Cradle of RNA

What were the conditions in which the first RNA molecules made their appearance on Earth? Even though we don't know the answer to this question, we may specify a certain number of conditions that are mandated by the knowledge that RNA did appear. In this exercise, I shall take as working assumption that the product was authentic RNA, not some hypothetical pre-RNA, and that the molecules' precursors were the same as they are today, namely ATP, GTP, CTP, and UTP. This hypothesis agrees with the congruence notion (see p. 17) and, more generally, obeys Occam's razor. With no valid reason to do so, why look elsewhere when what you need lies before your eyes?

A self-evident prerequisite of any protometabolic setting conducive to the formation of RNA molecules is the ready availability of the molecules' basic constituents: D-ribose, phosphate, and the four bases, adenine, guanine, cytosine, and uracil. Questions related to D-ribose and to phosphate have already been addressed (see Chapter 4). There remain the bases, which belong to the domain of nitrogen chemistry. There is good evidence, from a variety of sources, that small, nitrogenous radicals and molecules – in particular, ammonia (NH_3), hydrocyanic acid (HCN), and some of their derivatives – are abundantly produced by cosmic chemistry (Ehrenfreund et al., 2002). It is also well established that such substances can react fairly readily to form larger molecules, a number of which have indeed been observed in comets and meteorites or reproduced in the laboratory under plausible prebiotic conditions. Amino acids, mentioned before, are a case in point. Others are nitrogenous bases, including purines and pyrimidines. A whole literature exists on the topic (Schwartz, 1998).

At first sight, therefore, availability of building blocks does not seem to be a problem. What the real problem may rather have been is *embarras de richesse*. It is hardly likely that the chemistry involved could have produced just two pairs of bases that happened to be complementary and capable of joining to allow replication. Such felicitous coincidence between blind chemistry and far-reaching innovation smacks too much of intelligent design to be scientifically acceptable. As in the case of ribose (see "Why Adenosine?" in Chapter 4), the most likely hypothesis is that of the "dirty gemisch." If chemical conditions were such as to favor the synthesis of adenine, guanine, cytosine, and uracil, most likely they also furthered the formation of other purines and pyrimidines as well as other members of the vast heterocyclic family to which those bases belong. This family is represented in present-day living cells by a number of substances other than the canonical bases, often constituents of important coenzymes of presumably very ancient origin. Examples are nicotinamide, thiamine, pyridoxal, flavins, and pterins, to mention only a few. Some of these substances are even engaged in nucleotide-like combinations.

Our first master-word, therefore, in trying to define the cradle of RNA, is "gemisch." Lacking the exquisite specificities of metabolic reactions, the early chemistry was perforce "dirty." It made a much greater variety to start with than would be used in the long run.

Another master-word is energy. By definition, the cradle of RNA must have been continually traversed by a steady flow of energy capable of supporting the sealing of the bonds between sugars and bases to form nucleosides, between nucleosides and phosphate to form nucleotides, and between phosphates to form the pyrophosphate ends of the NTPs. The cauldron must have been cooking nonstop, to deliver an uninterrupted supply of ATP, GTP, CTP, and UTP, besides the many other products of the dirty gemisch.

According to earlier discussions, three possible, interconnected sources of energy could be considered within the context of the congruence notion. The most immediate and, at first sight, almost irreplaceable one is represented by inorganic pyrophosphate (or its polyphosphate polymers), which one would expect to be obligatorily required for the many phosphorylation reactions that dominated the chemistry involved. As we have seen, such substances could be produced under volcanic conditions (Yamagata et al., 1991). Note that the possibility that phosphate entered the system in a different way cannot be excluded (Arrhenius et al., 1993; Pitsch et al., 1995). Eventually, however, passage through pyrophosphate would seem necessary.

Barring a direct supply of energy-rich phosphate combinations, the next possible source, one step removed, would be thioesters, which we have seen can support the formation of pyrophosphate bonds. Mechanisms have been described whereby thioesters may indeed arise spontaneously, again in a volcanic setting.

Finally, there is the possibility, two steps removed, of electron transfers acting to generate pyrophosphate bonds, at least by way of thioesters. Protonmotive force is considered unlikely as a possible coupling mechanism because of its complex requirements. Potential electron donors being abundantly present among the building blocks, all that would be needed – not counting actual mechanisms – would be an appropriate electron acceptor, which would probably have been available fairly readily in a volcanic setting, or in any other.

All these conditions could not, of course, have created a hotbed for RNA assembly without a modicum of *catalytic* assistance. As we have seen, metal ions, clays and other mineral surfaces, iron–sulfur complexes, and my hypothetical multimers could all have been involved. Other factors, such as autocatalytic loops, could also have played a role. In the present state of our knowledge, nothing more can be said on this important topic, except that the reactions involved could hardly have been sufficiently specific to produce only those assemblages that turned out to be used by later life. What seems more likely is that classes of reactions took place. Sugars, not just D-ribose, interacted with nitrogenous bases, to produce a variety of compounds with the general structure of nucleosides, some of which could have acquired phosphate to turn into nucleotide-like combinations. Alternatively, sugars could have reacted first with mineral phosphate donors, to form a number of different phosphate esters, some of which then combined further with nitrogenous bases. Finally, some of the nucleotide-like combinations formed by these processes could have interacted with phosphate donors to give rise to diphosphate and triphosphate derivatives, including the canonical NDPs and NTPs but not excluding others. Such would appear to be, with our poor vision of the past, what Gerald Joyce, one of the leaders in the field, has called the "cluttered path to RNA" (Joyce, 2002).

The Birth of RNA

How RNA could possibly have emerged from the clutter without a "guiding hand" would baffle any chemist; it seems explainable only by *selection*, a process that presupposes *replication*. Imagine, as we almost have to, that the NTPs and other triphosphates present in the gemisch interacted to form internucleotide associations but that these were much more heterogeneous than RNAs, including other sugars besides ribose, other bases besides the four cardinal ones, and other linkages besides the 3′,5′-phosphodiester bonds that are found in RNA. Then, all we need to assume is that the mixture contained a few authentic RNA molecules and that these were somehow able to interact, in template fashion, with whatever mechanism was involved in their synthesis. Those two conditions being fulfilled, RNA molecules would progressively have become selectively replicated and amplified, up to a point where they became a dominant component of the mixture. In this way, RNA would have been born *together with replication* and as a result of it: the fruit of the very first true selection mechanism to take place in the development of life.

If so, on the basis of what criteria was RNA selected? The decisive factor would obviously have been base pairing, but this does not make A, G, C, and U mandatory. The gemisch could conceivably have contained other bases having appropriately complementary structures. Had such been the case, was chance or some hidden factor responsible for their rejection? According to a recent theoretical study, the choice of bases may not have been purely accidental but, rather, the outcome of a selective process influenced by resistance to error (Mac Donaill, 2003).

The choice of ribose raises a similar question. Was the selection of this sugar over other sugars fortuitous or due to some molecular properties that advantaged the ribose-containing molecules over the others? This question is of interest with respect to the possibility that RNA may have been preceded by some analogous substance of simpler structure (Joyce, 2002). Chirality is another factor. It is readily imagined that any heterochiral molecule would have been automatically barred from replication, as has indeed been found experimentally (Schmidt et al., 1997), making homochirality obligatory. The actual choice of chirality, however, could have been, as we have seen, the result of a frozen accident or, alternatively, of some physical bias (see Chapter 2). Whether the decision occurred at the precursor level or at that of the polynucleotides is, however, moot. Thus, the possibility cannot be excluded that L-RNA existed along with D-RNA and was subsequently eliminated by selection.

Given the hypothetical scenario just pictured, how did NTPs come to interact to produce the first RNA-like assemblages? If driven simply by chemistry, the first associations must have occurred essentially at random or, more correctly phrased, according to the intrinsic kinetic, steric, and energetic factors that governed the reactions, which factors may well have favored some associations and excluded others. In such a system, dinucleotides must have largely dominated, followed, in rapidly decreasing quantities, by trinucleotides, tetranucleotides, and higher-order oligonucleotides. Chains of some 35 nucleotides, which, as we shall see later, is a plausible size for the first true RNAs, could have been present in the mixture in only vanishingly small amounts, unless specially advantaged.

How longer chains could have been favored is not immediately obvious. Stabilization by binding to some surface or by folding into degradation-resistant conformations is possible, though such mechanisms, being purely passive, are not likely to account for the preferential selection of longer chains. The need seems inescapable for some autocatalytic process such that each lengthening step favors subsequent lengthening. Only in this way could the enormous kinetic obstacle to chain elongation be surmounted.

Such a situation could be achieved, perhaps with the help of multimers, by immobilization of the lengthening chain on a catalytic surface by forces that increase with chain length. A positively charged mineral, such as some clay–metal complex or pyrite, would bind the negatively charged nucleotide chains in this manner. As already mentioned, some success has been accomplished in the laboratory with such systems. Another possibility is autocatalysis. Let one end of the molecules (the 5′ end) include some configuration that favors the addition of nucleotide units to the other end (the 3′ end), let this activity increase with chain length, and chain lengthening will, indeed, be favored. Should this configuration at some stage turn out to be active also on other molecules, the lengthening could be extended to others. Hence the interest of investigators in possible RNA-assembling ribozymes (Johnston et al., 2001).

It must be kept in mind that any invoked catalytic mechanism must accommodate the participation of a template, for there can have been no emergence of true RNA molecules without replication. Furthermore, as will become clearer when we consider protein synthesis, the replication machinery must have been very sturdy, capable of functioning over very long periods of time.

The Ancestor of All RNAs

Leaving to further investigations the vexing problem of mechanism, what was the final product in this first stage of RNA history? This question has been addressed by the German chemist Manfred Eigen and his coworkers, who have devoted a considerable amount of work to trying to characterize the first RNA gene, the *Ur-Gen*. In a study based on both theoretical considerations and experimental observations (Eigen and Winkler-Oswatitsch, (1981), the conclusion was reached that the ancestral RNA was a GC-rich molecule, 50–100 nucleotides long, and most likely linked by direct filiation to present-day transfer RNAs, which, it will be remembered, are about 75 nucleotides long, on an average. Furthermore, the workers have found indications, from the comparison of the known sequences of transfer RNAs, that the common ancestor of these molecules could have consisted of two complementary chains, joined by a "hinge" allowing the chains to fold and join into a double-helical structure. These findings are relevant to what is known as the *sequence-space* problem.

The facts are well known. With four kinds of nucleotides available as building blocks, the number of possible distinct chains of *n* nucleotides

Table 7.1. *RNA sequence space.* Shown are the numbers N of different nucleotide sequences of length n and the total mass M of a set containing a single molecule of each species. Compare with Table 8.2. The formulas used for the calculations are

$$N = 4^n, \qquad M = \frac{N \times n \times 321}{6.02252 \times 10^{23}}$$

Length n (nucleotides)	Number of sequences N	Total mass M (g)
5	1.02×10^3	2.73×10^{-18}
10	1.05×10^6	5.59×10^{-15}
20	1.10×10^{12}	1.17×10^{-8}
30	1.15×10^{18}	1.84×10^{-2}
35	1.18×10^{21}	2.20×10
40	1.20×10^{24}	2.58×10^4
50	1.27×10^{30}	3.38×10^{10}
60	1.33×10^{36}	4.25×10^{16}
75	1.43×10^{45}	5.71×10^{25}
100	1.60×10^{60}	8.56×10^{40}
200	2.57×10^{120}	2.74×10^{123}

is 4^n. This number, the size of the sequence space, is small for short chains but, because of the exponential factor, quickly grows with increasing chain length, soon reaching totally unrealistic values (see Table 7.1). Most present-day RNAs, for example, occupy an infinitesimally minute area in an unimaginably immense sequence space, which fact, as will be seen in greater detail in the next chapter, is used as an argument by the defenders of intelligent design.

Leaving this question aside for the present, we may ask to what extent emerging life was able to explore the RNA sequence space as molecules of increasing length began to appear. Was the exploration extensive enough to allow selective *optimization*? Or was the final result only one in a number of possible ones, determined by chance factors unlikely to be repeatable?

As shown in Table 7.1, exploration of the sequence space could conceivably have been exhaustive up to chain lengths on the order of 35–40 nucleotides. With 35 nucleotides, a set containing a single sample of every possible sequence would have a mass of only 22 g. For a length of 40 nucleotides, the mass of such a set would be 26 kg still a manageable quantity. But beyond that length, things soon get out of hand. With 50 nucleotides, the mass of a complete set reaches 34 000 metric tons. For 75 nucleotides it amounts to one-hundredth of the mass of our planet. These calculations are of critical importance. They show that an ancestral RNA molecule 75 nucleotides long could not possibly have been arrived at by exhaustive exploration of the RNA sequence space, whereas this could readily have been true for a molecule half that long. In other words, if, as is believed by the Eigen school, the *Ur-Gen* consisted of two complementary units some 35 nucleotides long, it could have been the reproducible product of selective optimization, followed by duplication.

There is another interesting argument, also derived from the work of the Eigen school (Eigen and Schuster, 1977), suggesting that the length of replicating RNA chains could be subject to a natural limit. It comes from a theorem showing that the length of a replicable polymer cannot, in first approximation, exceed the reciprocal of the replication error rate. If it does, the molecule inevitably degenerates upon repeated replication. A maximum length of 75 nucleotides corresponds to a replication error rate of 1.3%, not an implausible value for a primitive mechanism.

Can more be said of the ancestral RNA? The answer to that question comes from experiments first performed in the 1960s by the late American molecular biologist Sol Spiegelman (1967) and since then repeated by many others. What the experiment has shown is that, when an RNA molecule is allowed to go through a large number of replication cycles, inevitably affected by repeated replication errors or accidents, the molecule automatically evolves toward a reproducible state characterized by an optimal combination of stability and replicability. This is the process of direct molecular selection, already mentioned in Chapter 3 (see the section "Congruence"). On the strength of such experimental findings, it may be surmised that the ancestral RNA progressively evolved into a stable form, or, more accurately, into what Eigen has called a *quasi-species*, consisting of the optimized product, or *master sequence*, accompanied by a continually changing cohort of mutant molecules arising from faulty replication and subject to natural selection according to environmental conditions.

What emerges from these admittedly very speculative considerations is: (1) that the road to RNA could have led to a reproducibly optimized molecule some 35–40 nucleotides long or, at most, twice that size, but containing a complementary duplex of the same information; (2) that further evolution of this molecule would have been determined initially by environmental influences on its stability and replicability; and (3) that this process was limited by the fidelity of the replication mechanism. These conclusions are crucially relevant to a hypothetical model that has acquired an almost mystical significance: the *RNA-world* model.

The RNA World

There are two facets to the RNA-world model. One was conjectured by Francis Crick as early as 1968 and is universally accepted. It was epitomized at the beginning of this chapter in two sentences: RNA came before DNA; RNA came before protein. Hardly anybody is inclined to quarrel with those assumptions, which rest on strong circumstantial evidence written into the fabric of present-day life.

The second facet of the RNA-world model followed the startling discovery, made in the early 1980s, independently by the Americans Thomas Cech and Sidney Altman, that RNA molecules can display catalytic activity. Cech coined the term "ribozyme" to designate catalytic RNAs and vigorously promoted the notion of "RNA as an enzyme" (Cech, 1986). This notion was adopted by another American, Walter Gilbert, of DNA-sequencing fame, and incorporated into his celebrated definition of an "RNA world" – a term he coined for the occasion – which he describes as a stage in the origin of life in which "RNA molecules and cofactors [were] a sufficient set of enzymes to carry out all the chemical reactions necessary for the first cellular structures" (Gilbert, 1986). Just one small word – *all* – converted a major discovery into a powerful hypothesis that has fired imaginations, galvanized research efforts, and inflamed passions in an astonishing manner, about which historians and sociologists of science of the future will have much to ponder.

Today, the RNA-world model has been popularized to such an extent that its speculative nature tends to be almost forgotten. Witness the following statement, which is only one of many in a similar vein: "Before biological coding emerged, the hypothesized RNA world, in which a single biopolymer performed both as genetic information and metabolically functional phenotype, forms one of the best supported models for the early evolution of life" (Freeland et al., 2003).

Having, in our attempted reconstruction of RNA history, reached the stage where the alleged RNA world was inaugurated, we may find it profitable to take an objective look at the model. Note that the occurrence of *an* RNA world is not in question. The point at issue is whether ribozymes ever came to catalyze "all the chemical reactions necessary for the first cellular structures" – or, if they never did so, to what extent they could have become involved in protometabolism before protein enzymes took charge.

A first question to be asked is whether the protometabolic "dirty cauldron" that spawned RNA could have continued to ensure this production until the advent of protein enzymes (see Chapter 3). Or did ribozymes begin to meddle with protometabolism already in the RNA world, to be followed later by protein enzymes? Absent clear proof of the participation of ribozymes, their involvement appears as a superfluous hypothesis. Protometabolism functioned before; why could it not have gone on doing so? Had it depended on extraordinary circumstances, not likely to be repeated or maintained for any length of time, the need for rescue could be invoked. But there is no reason to believe so. Protometabolism, as reconstructed, must have been solidly anchored in a robust set of local conditions (although, admittedly, reconstruction in the laboratory would be the only convincing argument in support of this view).

Even if not mandatory, the assistance of ribozymes could obviously have been very useful. So, what about the evidence? The most revealing clue is said to be the existence of a number of coenzymes with a nucleotide structure – NAD, NADP, FMN, and FAD are the most prominent examples – believed to be molecular "fossils" of corresponding ribozymes. The argument is impressive, but not decisive. As we have seen, the coenzymes could have arisen as parts of the gemisch, together with ATP and the other NTPs, which can hardly qualify as "fossils" of ribozymes, unless

RNA is taken to have arisen in a manner completely different from its present mode of synthesis.

Note that the coenzymes are not catalysts in the true sense of the word; they are cofactors. Present-day ribozymes show no evidence of including, or of ever having included, catalysts of the kind of group-transfer or electron-transfer reactions that might have been involved in protometabolism (the peptidyltransferase to be mentioned in the next chapter is an exception). There is, however, plenty of evidence that properly engineered RNA molecules *can* catalyze certain reactions of that kind (Joyce, 2002). Whether nature ever achieved in the past what clever engineers succeed in doing today is, of course, the key question. It would hardly be raised in the absence of positive evidence if the engineered material had not had the emblematic character of RNA and the technology used had not been borrowed from nature itself.

One phenomenon in which intervention of a ribozyme could have been of critical importance is RNA replication. As discussed above, catalysis of this process raises a serious problem. Among the conceivable solutions of this problem, the involvement, whether autocatalytically or otherwise, of an RNA replicase of RNA nature appears as an attractive possibility. This point is of great importance and justifies the search, already partly successful (Johnston et al., 2001), for RNA-replicating ribozymes. That such catalysts could have existed is rendered at least plausible by the fact that one of the two main functions in which ribozymes are involved today is RNA processing. It must, however, not be forgotten that the replicase at issue could have arisen only *after* RNA itself, whose origin remains unexplained (Joyce and Orgel, 1993).

A question raised by the ribozyme problem and directly related to our topic concerns the mechanism whereby catalytic RNAs could have been selected had chance produced the right mutation. This question has already been addressed in Chapter 3. We have seen that a process of molecular selection, of the kind alluded to above, is not readily visualized for a catalyst unless the catalytic activity happens to directly favor the selection of its own molecular support (as, for example, would occur in autocatalytic RNA replication). In a more general way, selection of a ribozyme would be expected to be indirect, like that of a protein enzyme, and to depend on the existence of competing protocells. The implication is that, if ribozymes did significantly take over protometabolic reactions, as surmised in the RNA-world model, the system must have been cellularized. This is not an objection, for cellularization seems to be a condition for the appearance of the first protein enzymes in any case. But the point will have to be kept in mind when the appearance of cells is discussed (see Chapter 10).

A last item of interest with respect to the RNA-world model concerns the length of the RNA molecules. As mentioned above, there is a serious possibility that the presumptive length of the ancestral RNA – about 75 nucleotides – represented an upper limit set by the error rate of primitive replication, in that longer molecules would inevitably have succumbed to repeated replication. An important implication of this hypothesis is that further RNA lengthening was conditioned on an improvement in the fidelity of RNA replication. If such was the case, the question arises whether this development had to await the appearance of protein replicases or was allowed to happen before, thanks to the participation of increasingly accurate RNA replicases of RNA nature.[4]

This question is related to the point, already discussed, of the possible involvement of a ribozyme in primitive RNA replication. All that can be said is that there seems to be no compelling evidence that RNA molecules of the assumed length – about 75 nucleotides – could not have functioned adequately until the advent of the first protein RNA replicases. As will be seen in the next chapter, the crucial development catalyzed by RNA molecules – with little doubt, here, as to its occurrence – is represented by protein synthesis. The main actors in this development are likely to be the ancestors of present-day transfer RNAs, that is, molecules about 75 nucleotides long. The other RNAs involved, presumably ancestral to ribosomal and messenger RNAs, need not have been longer, as far as can be seen. Thus, life could have effected the momentous transition from the RNA stage to the RNA–protein stage with a set of RNA molecules no more than about 75 nucleotides long.

In conclusion, except for the possible (though not so far demonstrated) involvement of an RNA replicase of RNA nature and, of course, for

---

[4] The participation of RNA-replicating ribozymes of increasing accuracy is a cornerstone of the theory proposed by Poole et al. (1999) (see footnote 2, in the next chapter).

the highly probable involvement of RNA molecules in the development of protein synthesis, there seems so far to be little evidence that ribozymes ever did fulfill the many catalytic functions implied in the original definition of the RNA-world model. This in no way invalidates the model as an imaginative hypothesis in search of verification (or falsification). Whether it will eventually be found to be true or not, its greatest value, as for many brilliant but not necessarily correct hypotheses of the past, lies in the experiments it has inspired and in the findings and achievements these experiments have generated. The model has been uncommonly fecund in this respect.

Irrespective of this issue, there is no denying that the advent of RNA was a true watershed in the development of life. It introduced replication, the essence of genetic *continuity*, and, with it, variation, competition, and selection, the bases of *evolution*. This we discern with the help of hindsight. It might not have been obvious to an observer at the time. Except for the slow introduction of RNA molecules, protometabolism may well have gone on largely unchanged, continuing to produce the same motley collection of molecules by the same variety of chemical reactions. The only significant change would have resulted from the appearance of ribozymes that altered the existing chemistry, a point of major uncertainty, as we have just seen. The true revolution initiated by RNA was the development of protein synthesis, without which the "RNA world," however defined, would have remained mostly barren. This will be the topic of the next chapter.

8 Proteins

Proteins are polypeptides, that is, long chains made of a large number of amino acids. A great variety of peptides exist in nature. Those that constitute proteins are made exclusively with twenty specific kinds of amino acids that are all α-amino acids [the carboxyl and amino groups are attached to the same (α) carbon atom] and, with the exception of glycine, which is not asymmetric, are of the L chiral variety (see Chapter 2):

$$
\begin{array}{c}
COOH \\
| \\
H_2N\text{-}C\text{-}H \\
| \\
R
\end{array}
$$

In proteins, as in other peptides, the amino acids are linked by peptide bonds, the outcome of the dehydrating condensation of the carboxyl group (COOH) of one with the amino group (NH$_2$) of another:

$$
\underset{R_1}{H_2N\text{-}CH\text{-}COOH} + \underset{R_2}{H_2N\text{-}CH\text{-}COOH}
$$

$$
\downarrow \tag{1}
$$

$$
\underset{R_1}{H_2N\text{-}CH\text{-}CO\text{-}NH}\underset{R_2}{\text{-}CH\text{-}COOH} + H_2O
$$

It is readily seen that the similarly linked attachment of additional amino acids to the dipeptide shown above can continue indefinitely, leading to a polypeptide molecule of increasing length, always having a free amino group (NH_2) at one end (N-terminal end, or *tail*) and a free carboxyl group (COOH) at the other (C-terminal end, or *head*). The length of protein molecules is usually on the order of several hundred amino acids but shows considerable variation. What gives each chain its individuality is the number of amino acids of which it is composed and, especially, their *sequence*, that is, the order in which they follow each other along the chain. This boils down to the sequence of R groups, because the backbone is the same for all chains, a repeat of the –NH–CH–CO– motif.

The R groups of amino acids are remarkable in the diversity of their physical and chemical properties. Some bear no electric charge; others are negatively or positively charged. Some attract water; others repel it. Some are chemically inert; others display various degrees of reactivity. These features make amino acids the most diverse and versatile of nature's building blocks and make their associations, by far, the most complex and multifunctional of all existing molecular structures, unmatched by any other biological constituent.

The sequence of amino acids determines the shape of a protein molecule and, thereby, all its physical and chemical properties. In relatively rare cases, the chains conserve their linear shape and join into various structural elements, such as fibers, plates, or trellises. Most often, the chains fold into complex, tridimensional bundles displaying on their surface intricate arrangements of R groups, which are largely responsible for the manner in which the molecules interact with their environment by way of binding sites, catalytic centers, and other sensitive areas. Most of the more complex components of living organisms, including cytoskeletal parts, enzymes, transporters and other energy transducers, receptors, regulators, antibodies, and a host of others, are proteins or have a protein foundation.

There is no doubt that the advent of proteins utterly changed the conditions of emerging life, representing a truly momentous watershed in its development. One would be tempted to call the event even more momentous than the advent of RNA, except that the proteins themselves are almost certainly products of RNA activity. Before examining how this may have happened, let us see how proteins are synthesized in present-day organisms. As with other events in the development of life, analyzing the present may yield useful hints concerning the past.

Protein Synthesis Today

Like all dehydrating condensations, that of amino acids requires energy. This is provided by ATP, by a sequential group-transfer mechanism that depends on the transfer of AMP [see reaction (6) in Chapter 4] to the carboxyl group of the amino acid, yielding an aminoacyl–AMP complex. The aminoacyl group is, however, not used directly for peptide bond synthesis; it is transferred to a special kind of RNA, called *transfer* RNA, or tRNA, that serves as carrier of this group:

$$\text{ATP} + \ \underset{\overset{|}{R}}{\text{HOOC-CH-NH}_2} \longrightarrow \underset{\overset{|}{R}}{\text{AMP-CO-CH-NH}_2} + \text{PP}_i \qquad (2)$$

$$\underset{\overset{|}{R}}{\text{AMP-CO-CH-NH}_2} + \text{tRNA-OH} \longrightarrow \text{AMP} + \underset{\overset{|}{R}}{\text{tRNA-O-CO-CH-NH}_2} \qquad (3)$$

$$\text{ATP} + \underset{\overset{|}{R}}{\text{HOOC-CH-NH}_2} + \text{tRNA-OH} \qquad (4)$$
$$\longrightarrow \underset{\overset{|}{R}}{\text{tRNA-O-CO-CH-NH}_2} + \text{AMP} + \text{PP}_i$$

With the amino acid represented by aa, reaction (4) can be summarized schematically as follows:

$$\text{ATP} + \text{aa} + \text{tRNA} \longrightarrow \text{aa-tRNA} + \text{AMP} + \text{PP}_i \qquad (5)$$

A bond is created between the amino acid and the tRNA at the expense of the splitting of ATP to AMP and inorganic pyrophosphate [see reaction (8) in Chapter 4]. Subsequently, the ATP-derived bond energy between the aminoacyl group and tRNA is used to create a peptide bond, though not by direct transfer of the activated aminoacyl group to the terminal NH_2 group of the growing polypeptide chain, as might be expected. Peptide chains grow not by their N-terminal tails but by their C-terminal heads.

This is best understood by considering first the initial reaction whereby a dipeptide is formed. In this reaction, the aminoacyl group of one aminoacyl–tRNA complex is transferred to another, yielding a dipeptidyl–tRNA complex:

$$aa_1\text{-}tRNA_1 + aa_2\text{-}tRNA_2 \longrightarrow aa_1\text{-}aa_2\text{-}tRNA_2 + tRNA_1 \qquad (6)$$

In the next step, the dipeptidyl group is transferred from its carrier tRNA to the terminal amino group of the next amino acid of the chain, reacting in tRNA-bound form, to yield a tRNA-bound tripeptidyl group:

$$aa_1\text{-}aa_2\text{-}tRNA_2 + aa_3\text{-}tRNA_3 \longrightarrow aa_1\text{-}aa_2\text{-}aa_3\text{-}tRNA_3 + tRNA_2 \qquad (7)$$

The chain goes on growing in this way by peptidyl transfer until the last amino acid has been added:

$$aa_1\text{-}aa_2\text{-}\cdots\text{-}aa_{n-1}\text{-}tRNA_{n-1} + aa_n\text{-}tRNA_n$$
$$\longrightarrow aa_1\text{-}aa_2\text{-}\cdots\text{-}aa_{n-1}\text{-}aa_n\text{-}tRNA_n + tRNA_{n-1} \qquad (8)$$

Finally, the completed chain is detached hydrolytically from its tRNA carrier:

$$aa_1\text{-}\cdots\text{-}aa_n\text{-}tRNA_n + H_2O \longrightarrow aa_1\text{-}\cdots\text{-}aa_n + tRNA_n \qquad (9)$$

From the energetic point of view, the closure of each peptide bond is supported by the splitting of ATP, by way of the aminoacyl–tRNA bond energy. The crucial point is that growing peptide chains remain attached to the tRNA of their C-terminal amino acid throughout their synthesis and that peptide bonds are sealed by *peptidyl transfer*, rather than by aminoacyl transfer.

In the entire living world, this process takes place identically on small particles, called *ribosomes*, made of roughly equal amounts of protein and RNA (ribosomal RNAs, or rRNAs). Significantly, the crucial peptidyl transfer step of protein synthesis is catalyzed not by a protein enzyme but by an *rRNA molecule*. This is one of the few cases in which a ribozyme acts in present-day life. As we shall see, it has profound implications.

A fundamental feature of protein synthesis is that the amino acids are not added in haphazard fashion. Their sequence is rigorously imposed by another kind of RNA, called *messenger* RNA, or mRNA, which is itself formed by transcription of a given stretch of DNA. This is how genes are expressed. The sequences of bases in DNA determine (by base pairing) the sequences of bases in the transcription products, the mRNAs, which themselves control the sequences of amino acids in proteins, the final agents of gene expression in most cases.

The DNA-derived information is written into the mRNA molecules in the form of successive triplets of bases, or *codons*, most of which specify a given amino acid (the others signal the end of the chain). The set of equivalences between amino acids and codons is called the *genetic code*.[1] It is the same for the vast majority of living organisms and represents the dictionary universally used in nature for the translation of nucleic acid language into protein language. We shall see later how this remarkable singularity may have emerged.

The coding action of mRNAs does not rest on direct interactions with the amino acids. As shown in Figure 8.1, what the mRNA codons interact with are complementary triplets, or *anticodons*, present in the amino-acid-carrying tRNA molecules. This interaction occurs by the universal mechanism of base pairing. Thus, mRNA molecules never "see" the amino acids; they "see" only the anticodons of the tRNA molecules that bear the amino acids. From this it follows that translation from the nucleic-acid language to the protein language takes place earlier, when amino acids are attached to their carrier tRNAs by reaction (5). The fidelity of translation depends critically on the accuracy of this process.

The enzymes that catalyze this crucial step are proteins called aminoacyl–tRNA synthetases. There are twenty of them, one for each

---

[1]   The term "genetic code" is increasingly used, in the media and even in certain scientific publications, to designate the genetic endowment, or genome, of an individual. This regrettable practice leads to dangerous confusion. To speak of an individual genetic code is nonsense. With rare exceptions, there is only a single genetic code for the entire living world.

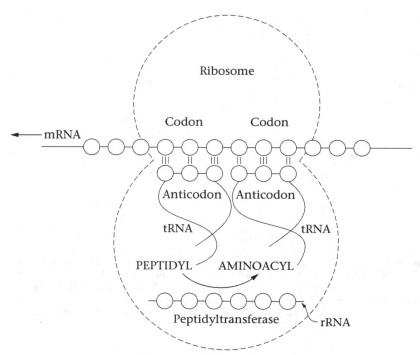

Figure 8.1. *Protein synthesis.* The mRNA runs through the ribosome (formed, as shown, of a small and a large subunit that are joined by the mRNA molecule), like a tape through a cassette reader. The "reading" is done by charged tRNA molecules. The diagram, which depicts an intermediate step in the process, shows two such molecules, kept side by side on the ribosome surface by base pairing between their anticodons and two adjacent codons of the mRNA. The tRNA on the left bears the growing peptide chain; that on the right, the amino acid destined to be added to the C-terminal head of this chain. The sites occupied by these tRNAs, called P (for peptide) and A (for amino acid), respectively, are strategically placed with respect to the catalytic site of the rRNA that functions as peptidyltransferase, so as to allow the transfer, by this ribozyme, of the growing peptide chain on the left to the aminoacyl group on the right. Following this transfer, the chain, lengthened by one unit, will now be attached to the tRNA on the right by means of the last-added amino acid, whereas the tRNA on the left, freed from its peptidyl group, will detach, leaving the P site empty. Once this has happened, the mRNA will move to the left by the length of one codon (with the help of energy provided by the splitting of GTP), dragging with it the tRNA bearing the lengthened peptide, which will occupy the P site. At the same time, the A site, which has become free and framed by a new codon, is ready to receive an aminoacylated tRNA containing an appropriate anticodon The situation depicted in the scheme will be reinstated, ready for the next elongation step. Two features of this mechanism are notable. First, the mRNA does not "see" the amino acid; it "sees" only the anticodon of the amino acid's carrier tRNA. Second, the main actors in the process are RNA molecules: mRNA, tRNAs, and the rRNA that acts as peptidyltransferase.

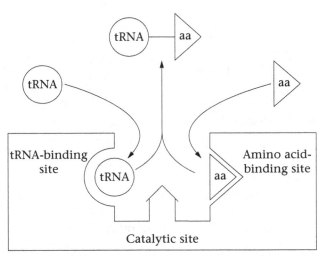

Figure 8.2. *Anatomy of a tRNA synthetase.* The enzyme has two binding sites, one for a tRNA, the other for an amino acid, situated in the vicinity of a catalytic site in such a way that the amino acid is attached to the tRNA. Not shown in this diagram is the participation of ATP according to reaction (5). Note that the correct joining of the amino acid with a cognate tRNA depends on the specificity of the two binding sites.

amino acid. As illustrated schematically in Figure 8.2, each enzyme has a binding site for the amino acid and another for the corresponding tRNA or tRNAs (there may be more than one). Note the exceptional importance of those twenty enzymes, which serve simultaneously in activation (energy) and in translation (information). They accomplish their energetic function by using ATP to join amino acids with tRNA molecules in a manner such that the energy of the created bond can serve to support peptide bond formation. Aminoacyl–tRNA synthetases carry out their informational function (see Figure 8.2) by selecting the substrates of these associations, so that amino acids always join specifically with tRNA molecules containing an appropriate anticodon. Remember that later selection by the mRNA depends exclusively on codon–anticodon interaction. The term *identity* designates the features whereby tRNA molecules are recognized by the aminoacyl–tRNA synthetases. Contrary to what one would expect, identity and anticodon coincide for only about half the enzymes. The other enzymes recognize their tRNAs by features different from the anticodon. This surprising peculiarity imposes an additional accuracy condition: identity and anticodon must be present together in the same tRNA molecule. That natural selection should have favored – or tolerated – this additional error risk is an intriguing fact. I shall come back to it.

The above account gives only the bare bones of what is obviously an extremely intricate process. But it should suffice for our purpose. Its main lesson is the central importance of RNA molecules (tRNAs, mRNAs, and rRNAs) in protein synthesis. This fact makes it highly probable that the first proteins arose from interactions among RNA molecules.

The Emergence of Proteins

It all started with certain RNA molecules reacting with certain amino acids. This much can be safely surmised, for it is difficult to see how else RNA molecules could have become involved with the stringing together of amino acids. Furthermore, it seems reasonable to assume that the postulated reaction led to the creation of a bond between the two molecules, prefiguring the attachment of amino acids to tRNAs. The direct relationship that may exist between present-day tRNAs and the *Ur-Gen*, the ancestor of all RNAs (Eigen and Winkler-Oswatitsch, 1981), further supports this hypothesis.

If a bond was made, energy would have been required. There are two simple possibilities. The energy could have been derived from a preexisting bond in the RNA molecule, of which one part would have been transferred to the amino acid, the other being released as leaving group. Alternatively, the amino acid could have reacted in activated form, for example, in combination with AMP, as happens today, or in some other form, such as a thioester.

A reaction of the kind postulated would probably not have taken place indiscriminately between any amino acid and any RNA. Most likely, some selection occurred, based on chemical affinities. It is tempting to assume that this is how the first amino acids destined for protein synthesis were selected from among those present in the gemisch. We have, indeed, seen (see the last section of Chapter 1) that the use of amino acids for protein synthesis does not reflect their alleged prebiotic abundance, indicating that some selective factor must have been at work. It does not seem likely, however, that all twenty proteinogenic amino acids were there right from the beginning. Some probably appeared and were selected later. By then, however, the tone may have been set, so that the choices of the future were limited by the commitments of the past. Several models have been proposed to account for a gradual recruitment of protein building blocks.

A key "decision" that may go back to those early days is that of chirality. It is an intriguing notion that RNA may be responsible for the selection of L- over D-amino acids. An even more intriguing corollary of this notion is the possibility that the chirality of RNA determined the chirality of the selected amino acids, implying that RNA made with L-ribose, if it had arisen, would have selected D-amino acids (de Duve, 2003). A recent report suggests that such could have been the case (Tamura and Schimmel, 2004).

An important question raised by these speculations is whether the primeval interactions between amino acids and their first RNA carriers were direct or indirect. A direct interaction would appear more likely for a primitive system. On the other hand, an indirect mechanism prefiguring the present-day situation, but with a ribozyme acting as aminoacyl–tRNA synthetase (see Figure 8.2), may deserve consideration. Note that selection of amino acid molecules by RNA molecules has to be presumed in either case.

As pointed out before, selection is a two-way process. If RNAs selected amino acids, the latter must, reciprocally, have selected the former. Chemical affinities are mutual. RNAs, however, were not there for the picking, as were the amino acids. They had to be actively selected on the strength of some advantage derived from their involvement. Before we consider this question, it may be useful to look first at what may have been the subsequent steps in the development of protein synthesis.

With only the present as guide to the past, the simplest hypothesis is to assume that the main features of protein synthesis were established right from the start. Take the scheme of Figure 8.1, remove the ribosomal infrastructure (which, as we know, includes many proteins that obviously could not have been present), add the prefix "proto" to the participating RNAs, and you have a model (Figure 8.3) of what could have been the

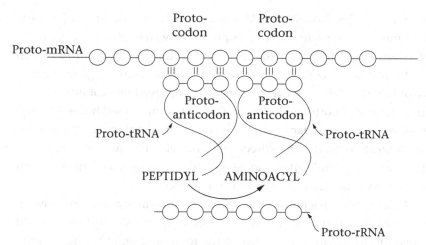

Figure 8.3. *Model of an all-RNA proto-ribosome.* Compare with Figure 8.1.

proto-ribosome, the ancestral system that inaugurated protein synthesis in the history of life.[2] Charged proto-tRNAs, immobilized side by side on a proto-mRNA molecule by base pairing between proto-anticodons and proto-codons, exchange aminoacyl and peptidyl groups with the help of a proto-rRNA catalyst. All this is highly conjectural, of course, but corresponds to what is about the simplest and most straightforward course of events that can be imagined to account for the development of RNA-dependent protein synthesis. Most workers who have thought about the question have come up with more or less similar solutions. I have myself adopted this model, with the additional twist that multimers may have played a role (de Duve, 1991).

If the proposed model is accepted, we must ask how the RNA molecules involved were selected. Remember that things presumably started with a single, ancestral RNA species or, more accurately stated, quasi-species (see

---

[2] In a scheme proposed by Poole et al. (1999), the proto-ribosome allegedly arose as an RNA-replicase system, not as a protein-synthesizing machinery. This scheme is based on a strong version of the RNA-world model, in which ribozymes, in addition to accomplishing many metabolic functions, acted as the principal agents of RNA replication and were the main participants in the process, to be mentioned below, whereby RNA molecules reached their maximum length thanks to progressive improvements in the fidelity of replication. According to this model, proteins came on the scene late, serving first as RNA-protecting "chaperones," and only later as enzymes.

the section "The Ancestor of All RNAs" in Chapter 7), from which the RNA initiators of protein synthesis presumably were picked out by natural selection. As already mentioned, selection may be direct or indirect. In the former case, the process is molecular and based on the greater stability of the molecules involved, or on their improved replicability, or on a combination of both qualities, resulting in the molecules' becoming progressively more abundant than those not favored in this way. In the case, much more frequent, of indirect selection, entities (cells or organisms) that possess the molecules emerge because this possession enables them to multiply faster than entities not similarly endowed.

As far as amino-acid-carrying RNAs are concerned, a molecular mechanism appears plausible. The presence of an attached amino acid could conceivably increase the stability of the molecules or, more importantly for a long-term effect, further their replication. It may be useful to remember, in this connection, that replication starts at what is known as the 3' end of the RNA template. If the amino acid had been attached to that end, as is the case for tRNAs, this feature could have facilitated the correct positioning of the template in relation to the replicating catalyst. It seems unlikely, however, that molecular selection could, by itself, have accounted for the selection of the other RNA molecules involved, including proto-mRNAs, the peptidyltransferase proto-rRNA, and, perhaps, the proto-aminoacyl–tRNA synthetases mentioned above. An indirect mechanism, dependent on the usefulness of the RNAs or on that of their products, seems almost mandatory, with, as prerequisite, the participation of a large number of competing protocells. Therefore, protocells must have been present already at the time protein synthesis was first developed.[3] This is an utter limit (de Duve, 2005). As we have seen, cellularization could even have taken place earlier.

Following the above line of reasoning, we are led to imagine a population of protocells, each endowed with a collection of replicating – and continually mutating – RNA molecules interacting with one another and

---

[3] I consider this point, which was already briefly made in Chapter 3, as of crucial importance. As will be seen in Chapter 14, it is highly relevant to our reconstruction of the LUCA, the last universal common ancestor of all known living beings.

with amino acids in a way conducive to the linking of the amino acids into peptide chains. In such a system, those protocells containing RNA molecules best able to interact in a manner useful to the survival and proliferation of their protocellular owners would be caused to emerge by natural selection. In my earlier speculations on this topic, the more or less random formation of small protein molecules was seen as providing the implied advantages, and the efficiency of their synthesis was seen as the criterion of selection.

An ingenious alternative proposal has been made by the Hong Kong scientist J. Tze-Fei Wong (1991) and further developed by the Italian Massimo Di Giulio (2003). According to this proposal, the agents of selection were not the proteins but the RNA-bound peptidyl groups. It is assumed that the participating RNA molecules exerted some useful catalytic activity and that this activity was enhanced by the attached peptidyl groups. This attractive hypothesis provides a functional and selective link between the RNA world and the later RNA–protein world, by way of what has been termed the PER (peptide-enhanced ribozyme) world. The hypothesis offers a plausible mechanism for the selection of molecules participating in the peptidyl-transfer reactions involved in this development. To the extent that peptide elongation increased the alleged ribozyme-enhancing activity of the peptidyl appendages, there would be continuing selective pressure favoring the RNA molecules most likely to participate in the formation of longer peptides. The same mechanism could even have applied to the other RNA molecules of the model if, as has been suggested, these molecules had themselves also been peptidylated. Note that, in this event, the proto-ribosome of the model (Figure 8.3) would have consisted of peptidyl–RNA molecules, prefiguring the protein–RNA associations of present-day ribosomes (Di Giulio, 2003).

The main weakness of the PER-world model is that it presupposes the existence of unspecified ribozymes, a standard feature of the classical RNA-world model but, as we have seen, supported by little objective evidence. It is, however, not certain that "*metabolic*" ribozymes would have been needed for the PER mechanism to operate. Perhaps the catalytic functions directly involved in protein synthesis, with the possible addition of the replicase activity considered in the preceding chapter, could provide the material for the alleged peptidyl enhancement.[4]

Further discussion of this fascinating topic must be left to the experts. A feature of the model in Figure 8.3 that the reader may have noted is that complementary base triplets – as opposed to groups of two or four bases, for example – are assumed to have ensured the immobilization of the proto-tRNAs on the proto-mRNA. Obviously inspired by hindsight, this feature could have been imposed right from the start by steric factors, thus serving as a straightforward, chemically determined pathway to the three-base genetic code. It is time for us to examine this crucial aspect of protein synthesis.

Translation and the Genetic Code

Ever since its elucidation in the early 1960s, and even before, the genetic code has invited an enormous amount of interest and speculation. The early history of the field has been covered in a previous work (de Duve, 1991). More recent information may be found in a comprehensive book on the subject by Osawa (1995), in Wong (1991) and Di Giulio (2003), already cited, and in Yarus (2000) and Wong and Xue (2002), among many others. For the present sketchy survey, the following points can be seen as particularly significant.

Universality of the Code. The vast majority of living organisms, be they microbes, plants, fungi, or animals, observe the same genetic code. There are a few exceptions, but these were almost certainly introduced later in evolution. They do not disprove the code's singularity; they simply show that the universal code is not immutable. I shall come back to this point.

Structure of the Code. The standard representation of the genetic code is shown in Table 8.1. To be noted first is that all of the 64 possible triplet combinations of three bases are used, with 61 specifying amino acids and

---

[4] An alternative, proposed by Poole et al. (1999), is that the proteins were selected by virtue of their ability to serve as chaperones for functional RNA molecules (see footnote 2, above).

Table 8.1. *The universal genetic code.* Codons marked * have a second function as initiators of translation. Codons marked † are terminators, or stop signs.

First letter	Second letter: U	C	A	G	Third letter
U	UUU Phe	UCU Ser	UAU Tyr	UGU Cys	U
U	UUC Phe	UCC Ser	UAC Tyr	UGC Cys	C
U	UUA Leu	UCA Ser	UAA OCHRE†	UGA OPAL†	A
U	UUG Leu	UCG Ser	UAG AMBER†	UGG Trp	G
C	CUU Leu	CCU Pro	CAU His	CGU Arg	U
C	CUC Leu	CCC Pro	CAC His	CGC Arg	C
C	CUA Leu	CCA Pro	CAA Gln	CGA Arg	A
C	CUG Leu	CCG Pro	CAG Gln	CGG Arg	G
A	AUU Ile	ACU Thr	AAU Asn	AGU Ser	U
A	AUC Ile	ACC Thr	AAC Asn	AGC Ser	C
A	AUA Ile	ACA Thr	AAA Lys	AGA Arg	A
A	AUG* Met	ACG Thr	AAG Lys	AGG Arg	G
G	GUU Val	GCU Ala	GAU Asp	GGU Gly	U
G	GUC Val	GCC Ala	GAC Asp	GGC Gly	C
G	GUA Val	GCA Ala	GAA Glu	GGA Gly	A
G	GUG* Val	GCG Ala	GAG Glu	GGG Gly	G

the other three acting as stop signals. Two amino acid codons double as initiation signals, by mechanisms that need not be examined here. There being more codons than amino acids, it follows necessarily that most amino acids are specified by more than one codon (up to six). There are only two exceptions to this rule: methionine and tryptophan. This peculiarity raises the question: Why so many synonyms? Why not a larger number of amino acids, for example, or more stop codons? As will be seen, there may be a historical reason for this.

Next to be noted is that the code is obviously not the result of a random assignment of codons to amino acids; it has a structure. Synonyms are grouped. In all cases, U and C are interchangeable in the third position. The same is mostly true for A and G as well, except for the two amino acids that have only one codon. For eight amino acids, all four bases are interchangeable in the third position. This relative lack of criticality is related to the fact that the pairing between anticodons and codons often enjoys a certain looseness in the third position, so that the same anticodon can pair with more than one codon. As a result of this phenomenon, known as *wobble*, the number of tRNAs and, therefore, of anticodons is smaller than the number of codons, usually lying somewhere between 35 and 45.

A more subtle feature of the code correlates the middle codon position with the affinity of the R group for water: U in that position characterizes the most hydrophobic (water-repellent) amino acids, followed by C for more moderately hydrophobic molecules, and by A or G for the most hydrophilic (water-attracting) ones. The presence of glycine in the latter group does not contradict that correlation. With only a hydrogen atom as R group, this amino acid may be said to escape classification based on water affinity.

An even more searching analysis has revealed that the structure of the genetic code seems to be such as to minimize the harmful consequences of point mutations, that is, of mutations leading to the replacement of one base by another in a message. When this happens, the relevant amino acid often either is not changed in the corresponding protein or is replaced in it by an amino acid with similar physical properties so that the protein remains functional. Computer simulations using hydrophobicity as a measure of kinship among amino acids indicate that less than one in a million possible codes could be superior in this respect (Freeland et al., 2003). As will be seen below, this feature suggests that the genetic code may be the product of selective optimization.

Birth of the Code. The seed of a genetic code is implicit in the scheme of Figure 8.3. If this scheme is a valid model of incipient protein synthesis, alignment of charged, peptidyl-exchanging proto-tRNAs along a proto-mRNA by the pairing of proto-anticodons with proto-codons would have taken place right from the start. For this feature to result in authentic coding, all proto-tRNA molecules possessing the same anticodon would have had to carry one and the same kind of amino acid. This condition would be most readily satisfied if proto-anticodons had been directly involved in the mechanism whereby RNAs first interacted with amino acids. As we have seen, this interaction is likely to have been specific. So-called stereochemical theories of the origin of the code have indeed been proposed, in particular by the Japanese investigator Mikio Shimizu, who has suggested, on the strength of molecular models, that anticodons could have served to create pockets accommodating the corresponding amino acids (Shimizu, 1982). This proposal does not seem to have been followed, however, and is hardly mentioned in recent discussions of the topic.

In fact, the most recent results suggest that codons, rather than anti-codons, could have played a role in RNA–amino-acid interactions. As reviewed by the American investigator Michael Yarus, a pioneer in this field, it has been found for three different amino acids (arginine, isoleucine, and tyrosine) that RNA molecules that naturally bind the amino acids, or have been engineered to display this property by *in vitro* selection, have an abnormally high frequency of codons specifying the amino acid situated in the vicinity of the amino-acid-binding site (Yarus, 2000). Interestingly, this kind of association would fit a proto-aminoacyl–tRNA synthetase of RNA nature, such as was mentioned above. Should such a ribozyme have bound its RNA substrate by means of a codon (see Figure 8.2), it would indeed have joined the amino acid to an RNA molecule bearing the appropriate anticodon.

Also possibly relevant to this abstruse problem is the fact that many aminoacyl–tRNA synthetases recognize their tRNA substrates by identity features different from the anticodons (see the end of the section "Protein Synthesis Today" above). It is not known how old or recent these peculiarities are. They could have been implicated as determinants of a "second genetic code," assumed to antedate the present one (de Duve, 1988).

A difficulty with any stereochemical theory of the origin of the genetic code is that all twenty proteinogenic amino acids were almost certainly not recruited for protein synthesis at the same time. It seems much more likely that primitive protein synthesis started with a small number of amino acids, possibly chosen among those that were most abundant in the primitive gemisch, and that additional amino acids were progressively introduced into the system after its main lineaments had already been set. Codons would, nevertheless, always have consisted of base triplets, imposed by the steric requirements of RNA alignment (see Figure 8.3). An ingenious model proposed by Eigen and

Winkler-Oswatitsch (1981) postulates three stages, in which first one, then two, and finally all three bases of the triplets become significant. Wong (1975) has proposed another interesting model linking the coevolutionary development of the genetic code and of amino acids to the metabolic reactions whereby new amino acids may have arisen from the primeval ones.

A major complication that besets all attempts at reconstructing the history of the genetic code from currently available information is that the code seems to be, at least in part, the product of a selection process. A few words about this possibility, which raises fascinating questions, are in order.

Selection of the Code. The term "frozen accident" (mechanism 5) was invented by Crick to characterize the genetic code. It turned out to be a double misnomer. The code is not accidental. And it is not frozen, to the extent that a few deviations from the universal code have taken place in the course of evolution. Crick thought this possibility to be ruled out because any mutation changing the significance of a codon would affect so many proteins that at least one indispensable protein was bound to suffer a fatal modification. It is significant in this respect that code changes have been found mostly in mitochondrial genomes, which, being very small, have a better chance than nuclear genomes not to include genes for potential protein targets of the mutations. Even a number of microorganisms, however, have now been found to deviate from the universal genetic code. As reviewed by Osawa (1995), at least two possible mechanisms exist whereby such changes could have occurred without harmful consequences.

Such episodic occurrences have nothing to do, however, with what is being considered here. What we are dealing with is an evolutionary process in which enough different codes are put to the test of natural selection for near-optimization to occur. Let us first consider the mutations involved in such a process. For the code to be modified, either the anticodon of an RNA carrier or the nature of its attached amino acid must be changed. Possible targets are the anticodon itself or other molecular features of the carrier RNA involved in its interaction with the amino acid or with the catalyst that links it to the amino acid (aminoacyl–tRNA synthetase in modern systems). Alternatively, the mutations could affect a binding site in this catalyst, so that it recognizes another tRNA or another amino acid (see Figure 8.2). Clearly, a large number of such mutations must have taken place for the singularity to emerge from a true selective bottleneck (mechanism 2).

The problem is that, for mutant protocells of this kind to be subject to selection, they must be viable and able to compete with their unmutated congeners on the strength of their long-term ability to withstand the harmful consequences of point mutations. This is a highly restrictive condition, because code changes usually have drastic consequences. They affect all the proteins in which one or more molecules of an amino acid are replaced by another as a result of the new coding. As mentioned above, in modern systems such a situation is almost invariably lethal. For this reason, tolerated mutations would be expected to be very rare. Yet, we have just seen that they must have been numerous to allow selective optimization of protocells on the basis of resistance to mutations.

There is thus an apparent antinomy between the two conditions required for the alleged optimization of the genetic code by natural selection. Once again, intelligent design looms on the horizon.

Not so fast! There is a possible solution to the conundrum. Its name is *redundancy*. In DNA genomes, this trait is most often reduced to mere duplication (e.g., paralogy, diploidy), with consequences that experience has shown can be highly beneficial. But, in an RNA genome, genes and messengers are one and the same molecules, usually present in a number of copies. In such a system, innumerable mutations may take place without lethal effects. If one gene molecule and its translation product are disabled, many other unharmed molecules remain to carry on the function involved. There are plenty of opportunities for experimentation, even by further modifications of disabled molecules, which would continue to be synthesized. This, incidentally, is a powerful argument for assuming that optimization of the code occurred *very early* in the development of protein synthesis, at a stage when the process was still dominated by RNA molecules and the synthesized proteins were still very short and few in number.

A system of this sort has not, to my knowledge, been studied by theoretical modeling. I can only share my intuitive suspicion that it would allow some kind of coevolutionary development in which code and functional products would be jointly subjected to selection. It seems possible that this sort of genetic gemisch could have offered enough opportunities to blind selection for a near-optimal code to emerge.[5]

The Growth of Proteins

If there is any truth in the preceding considerations, the first proteins must have appeared long before the genetic code had achieved its present structure and must have played a key role in the selective process whereby this structure was attained. This implies that even the primitive proteins produced at this early stage must already have affected by their properties the ability of their protocells to survive and multiply. Catalysis was likely the most important such property. Thus, the mechanism, mentioned in Chapter 3, whereby protein enzymes progressively took over protometabolism must have started at that time.

The first proteins were almost certainly very short. If, as indicated in the preceding chapter, the first RNA genes were on the order of 75 nucleotides long, their translation products would, at the most, have been 25 amino acids long, possibly even shorter if, as can be expected, the molecules were not translated over their whole length. It thus appears that protein molecules this short can display enzyme-like catalytic activities. This point, which is sometimes disputed, is of some importance with respect to the possible catalytic activities of the multimers proposed in Chapter 3.

Many present-day proteins contain several hundred amino acids. Most likely, these long molecules arose by gradual lengthening of the first products of protein synthesis, a consequence itself of the gradual lengthening of the coding RNA molecules. As seen in the preceding chapter, the maximum length of an RNA molecule is probably set by the fidelity of its replication.[6] In line with this requirement, the pace-setting event in protein

---

[5] A point that may have to be taken into account is the historical factor. It is possible that the genetic code owes its present structure at least partly to the manner in which amino acids and codons were progressively recruited, rather than to selective optimization.

[6] In first approximation, the length of an information-bearing polymer cannot exceed the reciprocal of the replication error rate. If it does, the molecule irreversibly loses its information upon repeated replication (Eigen and Schuster, 1977).

evolution could well have been the progressive improvement in the accuracy of the replicating catalyst, presumably a protein enzyme that replaced the primitive replicase.[7] Given this condition, longer RNAs could survive repeated replication, leading to longer proteins, among which, presumably, more efficient enzymes would be present, enough for selection to favor the protocells in which the changes took place. Thus, proteins would have climbed from one level of complexity to another, bootstrapped, so to speak, by the progress achieved by one of them, RNA replicase.

As to the mechanism of lengthening, what is known of the structure of present-day proteins strongly suggests a combinatorial process in which existing RNA stretches were associated in various manners, such that the corresponding translation products consisted of combinations of existing peptide chains. Proteins, indeed, show evidence of such an origin. They often consist of domains, or motifs, that can be recognized, diversely combined, in a number of different molecules. Needless to say, a variety of mutations would have been superimposed on this combinatorial shuffling.

Interestingly, in present-day cells, especially of higher organisms, the primary RNA transcription products of DNA genes frequently undergo a process, known as *splicing*, in which certain segments, called *introns*, are excised by hydrolytic cutting, and the remaining segments, called *exons*, are respliced together to form the mature messenger. Catalytic RNAs play a dominant role in this processing, which, next to protein synthesis, represents the second major involvement of ribozymes in today's life. It is tempting to assume that this activity goes back to the early days when RNA genes were first spliced into longer genes.

The notion, almost imposed by what is known, that proteins started as short molecules that attained their present size by progressive lengthening is highly relevant to an argument, known as the sequence space paradox, that is often alleged by defenders of intelligent design against the view that life originated naturally. The paradox rests on the kind of calculation illustrated for RNA in the preceding chapter (see Table 7.1) and now repeated for proteins.

---

[7] The argument would remain valid if, as is sometimes implied in the RNA-world model, the replicase had been of RNA nature (see footnote 2). At some stage, however, a protein catalyst must have entered the scene.

Table 8.2. *Protein sequence space.* Shown are the numbers N of different amino-acid sequences of length n and the total mass M of a set containing a single molecule of each species. Compare with Chapter 7, Table 7.1. The formulas used for the calculations are

$$N = 20^n, \qquad M = \frac{N \times n \times 134}{6.02252 \times 10^{23}}$$

Mass of Earth = 6×10^{27} g; mass of Galaxy = 8×10^{44} g; mass of universe = 1.7×10^{56} g.

Length n (amino acids)	Number of sequences, N	Total mass M (g)
10	1.02×10^{13}	2.28×10^{-8}
20	1.05×10^{26}	4.67×10^{5}
25	3.36×10^{32}	1.87×10^{12}
30	1.07×10^{39}	7.17×10^{18}
40	1.10×10^{52}	9.79×10^{31} (1 600 × Earth)
50	1.13×10^{65}	1.25×10^{45} (Galaxy)
100	10^{130}	2.22×10^{110} (10^{54} universes)

Consider a protein chain constructed, as proteins are today, with twenty different kinds of amino acids. For the first amino acid, twenty possibilities exist. If a second amino acid is added, the number of possible resulting dipeptides is 20^2, or 400. It is 20^3, or 8000, for tripeptides, 20^4, or 160 000, for tetrapeptides, in general 20^n for a molecule of n amino acids. This number defines the protein sequence space. Some calculated sequence–space values are shown in Table 8.2 and converted, for easy visualization, into the mass of a mixture containing a single representative molecule of every possible sequence. It is readily seen that the protein sequence space quickly reaches unimaginably high, materially impossible values. Even at the 50-amino-acid level, a collection of single representatives of all possible sequences would reach the mass of our Galaxy. For molecules containing several hundred amino acids, even the most gigantic "multiverse" conceived by cosmologists would not suffice. It is on the strength of such figures that the defenders of intelligent design claim that emerging life could not possibly have arrived at its proteins by a random exploration of the sequence space; it must have been guided.

But that is not how things happened. Life, as we have seen, most likely started with very short proteins, less than 25 amino acids long. Let us take an average length of 20 amino acids. We see (Table 8.2) that the corresponding sequence space contains 10^{26} possibilities, assuming, which is far from certain, that all twenty proteinogenic amino acids were already available for protein synthesis at that time. This number is large, but not inordinately so. The same number of protocells, assumed to be the size of a modern bacterium (1 μm in diameter), would, if they occupied one-thousandth of the volume, fit in a moderate-size lake. Considering, further, that the postulated selection process probably involved a large number of successive protocell generations and that the mutation rate of the RNA genes must have been very high, it is not inconceivable that the early evolving protocells had the opportunity to subject to natural selection most of the possible sequences contained in the sequence space, with a state near optimization as outcome. This is all the more likely in that this phase probably coincided with a good part of the process of code selection (see above) and must, therefore, have lasted a very long time.

There is a weakness in the above calculation. It is not the protein sequence space that was explored, but the RNA sequence space, which is ten orders of magnitude larger: 4^{60}, or 1.3×10^{36}, for a chain of 60 nucleotides, the minimum length to code for a 20-amino-acid protein. All the oceans in the world would not suffice to house this number of protocells. The difference between the two spaces is obviously due to the fact that the same amino acid is most often represented by more than one codon in the genetic code. Consequently, the same protein may be encoded by a huge number of different RNA sequences. This means that exploration of the protein space by way of the RNA space, though less exhaustive than it would be if it occurred directly, would still be very extensive in most cases. Note that if a smaller number of amino acids had been used for protein synthesis at that time, as seems likely, complete exploration of the spaces would have been considerably easier.

Some of the sequences put on trial at the stage considered must have been useful to the proliferation of their protocellular owners and thereby become selected.[8] As we saw in Chapter 3, a number of primitive enzymes and ribozymes may have emerged in that way. Two activities, in particular, were critical. One was the ability to replicate RNA more accurately, without which lengthening of the molecules would not have been possible. The other, if it did not already exist, was the ability to splice RNA segments together (ligase activity). As mentioned above, this activity was probably the main instrument of RNA lengthening.

With those two conditions fulfilled, longer molecules could have been tested, with as new size limit the length set by the error rate of the improved replicating catalyst. The sequence space available for this kind of exploration would, however, not have comprised all the possible sequences of this size (Table 8.2), but only those that could be constructed by combination of the RNAs retained in the first round and their mutants. Housed in protocells selected by virtue of the advantages they derived from their RNA genes, the molecules available for combination would perforce have been present in a limited number of varieties, small enough to allow subjecting the vast majority of combinatorial possibilities to natural selection. The same succession of combination followed by selection may be expected to take place with every improvement in replication accuracy. The result is selective optimization without the help of a guide. Figure 8.4. shows a highly simplified representation of such a process.

The above description, needless to say, is an extremely gross approximation of what may have taken place in reality. But it should suffice to illustrate how proteins could have arisen and developed by a natural

[8] An interesting question concerns the manner in which useful mutations were allowed to spread. If transfer was exclusively vertical, that is, generational, useful mutations must have followed each other serially in the course of successive selection events. If, as several authors have recently claimed (see Chapter 14), horizontal gene transfer took place at a high rate at that time, advantageous innovations could rapidly have been shared by neighboring protocellular populations, thus speeding up evolution. Such sharing of genes cannot, however, have been so extensive as to interfere with Darwinian competition, an essential condition of selection.

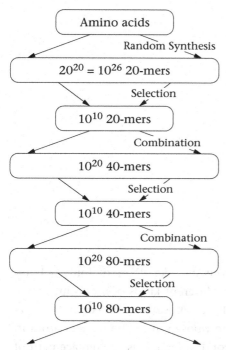

Figure 8.4. *Principle of the modular lengthening of proteins.* The diagram shows in highly schematic form how successive cycles of modular combination (by way of RNA genes) and selection may lead to very long protein molecules by a pathway that allows exhaustive exploration of the available sequence space at each stage.

process depending, once again, on the testing by natural selection of a sufficiently wide range of possibilities to allow something close to an optimal outcome to emerge. It is noteworthy that this same conclusion has been reached, on objective grounds, whether RNA or protein sequences are considered.

9 DNA

DNA, or deoxyribonucleic acid, consists, like RNA, of long polynucleotide chains assembled from four different kinds of mononucleotide units. There are two differences. First, ribose is replaced in DNA by 2-deoxyribose, which corresponds to ribose minus the oxygen atom in position 2 of the molecule. The second difference is the replacement of uracil by thymine, which is uracil with an added methyl group (CH_3). This modification does not affect the ability of the molecule to pair with adenine.

In contrast to RNA, almost all the DNA in nature is in double-stranded form, known worldwide as the double helix. Single-stranded DNA is found only in a few rare viruses. Intriguingly, the exact opposite occurs with RNA, which is almost always single-stranded, except in some rare viruses.

The replication of DNA takes place in the same way as that of RNA. The precursors are the deoxyribonucleotide triphosphates, which are distinguished from their ribonucleotide congeners by the prefix d: dATP, dGTP, dCTP, and dTTP. Base pairing rules the process, as for RNA, except that A pairs with T instead of with U. The double-stranded nature of the template does, however, introduce a number of complications, which are solved by the participation of unwinding and rewinding enzymes (helicases, gyrases, topoisomerases). In addition, because the two DNA strands have opposite polarities and replication obligatorily follows the template in only one direction (from its 3' end to its 5' end), one strand is replicated by a continuous process, generating what is called the leading strand, whereas the other strand is replicated discontinuously into a series of short stretches (Okazaki fragments) that are assembled in a direction opposite to the direction of assembly of the leading strand and are later joined together by a ligase, to form the lagging strand. Finally, DNA replication is further controlled by "proofreading" enzymes in such a way that most mismatches are detected and suppressed before they become part of the structure. Thanks to this control, the fidelity of DNA replication is orders of magnitude greater than that of RNA replication, amounting to the fantastically low error rate of one wrongly inserted deoxynucleotide in about one billion.[1]

The Birth of DNA

In nature, the building blocks of DNA arise from those of RNA. Deoxyribose is formed by the reductive deoxygenation of ribose, and thymine by the methylation of uracil. Both processes involve nucleotide derivatives, which are most often the nucleoside diphosphates for the reduction of ribose and dUMP for uracil methylation (to dTMP). It is tempting to assume that, in the origin of life, deoxyribose likewise arose from ribose, and thymine from uracil, although possibly by different mechanisms. At first sight, one could imagine that such simple reactions might have occurred early in protometabolism and that the DNA constituents already existed in the primeval gemisch. Note, however, that the apparent simplicity of the reactions is deceptive. In the present world, the conversion of ribose to deoxyribose and that of uracil to thymine are catalyzed by fairly complex enzyme systems. The possibility cannot be excluded, therefore, that the chemical makings of DNA appeared late, when a number of enzymes had already been developed.

Once deoxyribonucleoside triphosphates (dNTPs) were available, it is likely that their assembly into DNA-like chains would have followed

---

[1] This astonishing figure is equivalent to the performance of a typist copying the *Concise Oxford Dictionary* some fifty times, making a single mistake.

fairly rapidly, at least in a world where RNA was already being synthesized and replicated. Such an event could have required only minor innovations, if any. The primitive catalyst involved in RNA replication could very well, without much change, have constructed a complementary DNA molecule, instead of a complementary RNA molecule, on an RNA template (reverse transcription).[2] Once DNA molecules were on the scene, the same catalyst could, again without much change, have catalyzed the replication of these molecules as well as their transcription to RNA.

We are thus led to visualize a situation, set forth by the appearance of deoxyribose and thymine, in which the genetic information could have flowed fairly freely and reversibly between RNA and DNA. The question arises as to what selective forces favored the deviation of the two nucleic acids into entirely different functional directions.

Why DNA?

In order to address this question, let us look at a typical present-day example, the bacterium *Escherichia coli*, a common denizen of the human gut. It has some four billion years of evolution behind it, but the main lines of its genetic organization may well go back to the common ancestor of all life on Earth.

In *E. coli*, all the DNA of the cell is contained in a single, circular, double-stranded molecule, or chromosome, of about three million base pairs. More than one thousand distinct genes follow each other in this giant molecule, separated by untranscribed stretches, many of which have regulatory functions. Shortly before cell division, replication of this structure is initiated, starting at a specific site, called the origin of replication, and proceeding all around the template circle to finally join the starting point by the back, where a last bond closes the new circle and frees it from the template. At cell division, each daughter cell takes one of the two circles.

---

[2] This reaction does not take place in normal cells. It occurs in cells infected by certain RNA viruses that reproduce their genomes by way of DNA, thanks to a virus-specific enzyme named *reverse transcriptase*. Called *retroviruses*, these viruses include the infamous agent of the acquired immunodeficiency syndrome (AIDS) and a number of cancer-causing viruses.

In between two replication cycles, the DNA is subject to transcription into RNA molecules, which, except for a few functional RNAs, such as those involved in protein synthesis, are further translated into proteins. A key property of transcription is regulated selectivity. Each gene or, sometimes, group of neighboring genes (operon) is transcribed at a specific rate, which depends in complex fashion on internal and external conditions. This regulation is accomplished by special proteins, called transcription factors, that bind to specific, strategically situated sites of the DNA and there turn on or off, or enhance or slow down, the transcription of the genes governed by the sites. Still relatively simple in *E. coli* and other bacteria, transcriptional regulation becomes increasingly complex in higher organisms, where, among other functions, it plays a fundamental role in cell differentiation and embryological development.

Contrast this situation with that supposed to have existed at an early stage in the development of life, when there was no DNA, and RNA served both as replicable repository of genetic information and as agent in the expression of that information. The advantages brought about by DNA are staggering.[3] Replication has become an orderly affair, in which single copies of every gene are made together at a precisely set time. Transcription, on the other hand, is entirely dissociated from replication and is regulated at the single-gene level.

This is not all. In present-day cells, transcription most often involves only one of the two DNA strands, producing a single-stranded RNA free to adopt highly complex and functionally versatile configurations, in contrast with DNA, which is shackled into the rigid, unreactive double-stranded conformation. Things must have been very different when RNA played both roles. In fact, it is difficult to imagine a situation in which

---

[3] The primary advantage of DNA over RNA as a genetic material is frequently stated to be the greater chemical stability of DNA (see, for example, Joyce, 2002). The presence of a hydroxyl group in position $2'$ of ribose does indeed increase the sensitivity of the adjacent $3'-5'$ phosphodiester bond to hydrolysis. But this is true mainly in alkaline media. DNA is the more fragile molecule in acid media, where it readily loses its purine bases. This trait is exploited in the colorimetric Feulgen reaction for the detection of DNA. What seems to me important in DNA is not so much its chemical nature but the functions it fulfills. It allows replication to be synchronized and to be dissociated from transcription, with, as added benefit, the possibility of selective transcription control.

a number of short RNA strands could be replicated continually in more or less haphazard fashion without becoming irremediably entangled into double-stranded associations with their replication products and therefore unable to accomplish any function, such as initiating protein synthesis. The problem is very real, for RNA strands complementary to mRNAs (antisense RNAs) are being used successfully to inhibit the expression of certain specific genes. Yet, if there is any truth in the widely accepted model of an RNA-dominated world, nascent life must have traversed such a stage without mishap. How it managed to do so poses an intriguing problem for future research.

Directly replicating,[4] single-stranded RNA viruses give us a hint as to how things could have happened. When such a virus infects a cell, its RNA is first replicated into a double-stranded structure called the replicative form. The enzyme that catalyzes this process (RNA replicase) is not provided by the host cell; it is encoded by the viral RNA. There are two possibilities. In one (polio virus, for example), the viral RNA, known as a plus strand, acts as messenger. It is translated by the cell to give the replicase, which then starts to function. In the other case (measles, for example), the viral RNA is a minus strand and must first be replicated to the complementary plus strand for translation into the enzyme. Such viruses carry the required replicase to start the job. In both cases, a double-stranded replicative form serves as template for replication and for the preferential formation of plus strands serving in the translational synthesis of viral proteins. We thus have a situation resembling the normal DNA–RNA relationship, with double-stranded RNA playing the role of DNA and single-stranded RNA that of RNA. Perhaps an arrangement of this kind emerged early, creating some order in what would have been utter chaos and retained by selection thanks to this advantage.

When DNA?

Whichever way DNA first appeared, the functional separation allowed by this event must have been a tremendous improvement, favored by very

---

[4] Indirectly replicating RNA viruses are the retroviruses mentioned in footnote 2 of this chapter.

strong selective forces. One would be tempted to assume that DNA took over as soon as its appearance became chemically possible. One would have thought so, but for the possibility, raised in the preceding chapter, that the high degree of redundancy provided by bifunctional RNAs may have permitted the early selections that led to optimization of the genetic code and to the appearance of the first enzymes. We may thus visualize a stage of some confusion, in which redundancy (RNA) and centralization (DNA) vied on the strength of the selective benefits each conferred upon the protocells involved.

As we know, DNA ended up the winner. But this may have happened only after emerging life had developed its final genetic code as well as its first enzymes. Even the process of gene lengthening mentioned in the preceding chapter could still have involved RNA genes, as is suggested by the role of ribozymes in RNA processing today. This argument, however, is weak, because enzymes that cut and ligate DNA stretches also exist. In any event, it is obvious that DNA did not suddenly come to reorganize genetic relationships as by the stroke of a magic wand. There must have been a long intermediary period during which DNA, helped by natural selection, progressively took over the replicative storage function of RNA, leaving the latter to continue providing ribozymes and, especially, messengers for the synthesis of proteins.

10 Membranes

Cells owe their individuality to their being entirely surrounded by a thin molecular "skin," or membrane. Let this outer envelope be torn, and the semifluid cellular content leaks out; the cell dies. In addition, the inside of eukaryotic cells (see Chapter 15) and of some prokaryotic cells is further partitioned into a number of membrane-bounded compartments. All biological membranes are built according to the same blueprint, yet another of the remarkable chemical singularities that characterize life.

The Universal Fabric of Membranes

The basic fabric of all membranes is the *lipid bilayer*, a bimolecular layer consisting of molecules typically made of a three-carbon core or skeleton to which are attached, on one side, two long tails composed exclusively of carbon and hydrogen and, on the other, a head containing a variety of substances that share the property of being electrically charged or polarized:

$$\begin{array}{l} CH_3 \diagdown\!\diagup\!\diagdown\!\diagup\!\diagdown\!\diagup \boxed{\begin{array}{l} C \\ | \\ C \\ | \\ C \end{array}} \\ CH_3 \diagdown\!\diagup\!\diagdown\!\diagup\!\diagdown\!\diagup \\ C\!-\!\!\text{(head)} \end{array} \qquad (1)$$

Such molecules are known as *amphiphilic* (literally, having two loves), by which is meant that their two ends have very different affinities for water. The head is *hydrophilic* (water-loving); it tends to bind water molecules by electrostatic attractions, that is, those created by the so-called Coulomb forces, whereby electric charges of opposite sign attract each other (and charges of the same sign repel each other). This joining stems from the fact that the water molecule, although uncharged, is electrically polarized, with the oxygen atom protruding on one side and forming the negative pole, and the two hydrogen atoms creating two neighboring, positive poles on the other. Note that this structure makes the water molecule itself hydrophilic. Water molecules tend to bunch together in a disorderly network, held by electrostatic attractions linking the negative pole of one molecule to a positive pole of another, and so on. This is what makes water liquid. But for this property, water would be gaseous at even very low temperatures.

The tails of the bilayer constituents are *hydrophobic* (water-hating). Actually, this denomination is incorrect. Hydrophobic groups do not actually repel water. They are indifferent to it, leaving its molecules free to cluster by electrostatic attractions. At the same time, hydrophobic groups are attracted to each other – they are often called *lipophilic* (fat-loving) for this reason – by weak forces known as van der Waals forces. This, basically, is why oil and water don't mix. Oil molecules stick together by van der Waals attractions, water molecules by Coulomb attractions, with essentially no mutual interference. It is said that they exclude each other.

When shaken with water, amphiphilic molecules spontaneously organize in a manner that best satisfies the conflicting requirements of their hydrophilic heads and of their hydrophobic tails (by which is meant, of course, that they adopt the configuration of lowest energy). Micelles, foams, and bubbles are structures of this kind. The lipid bilayer is the favored configuration with the constituents of biological membranes. In this structure, two monomolecular layers join by their hydrophobic faces, creating an inner, oily film, lined on both sides by the hydrophilic heads in contact with water:

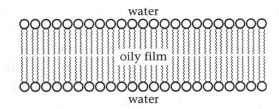

As will be mentioned below, such structures automatically close into saclike vesicles, so that the bilayer separates two watery milieus, occupying the inside and the outside of the vesicle.

By far the most common membrane constituents are phospholipids, in which the head [see formula (1)] is made of a negatively charged phosphate group, which, in a number of cases, is itself linked with a positively charged basic molecule, such as choline or ethanolamine. Also found in many membranes are glycolipids, in which the head consists of a carbohydrate derivative, sometimes associated with an acidic group, such as sialate or sulfate.

The three-carbon core of the molecules is most often the trialcohol glycerol, except in sphingolipids, in which the long-chain amino alcohol sphingosine provides both the core and the upper hydrophobic tail.

In the membranes of all eukaryotic cells and of all prokaryotes classified as Bacteria, the hydrophobic tails usually belong to long-chain fatty acids linked to the core by *ester* bonds (or by an amide bond, when linked to sphingosine, which, as mentioned above, also provides one of the hydrophobic tails). In Archaea, on the other hand, the hydrophobic tails belong to polyisoprenoid alcohols attached to the core by *ether* bonds (Kates, 1992). This is a major difference between the two prokaryotic domains. So is the fact that the central glycerol molecule has opposite chiralities in the two kinds of lipids (see Figure 10.1).

Biological membranes owe a number of distinctive properties to the physical characteristics of the lipid bilayers that form these membranes. One such property is *flexibility*. Because the forces that hold the hydrophobic tails of the bilayer molecules together are weak, these molecules can readily slide along each other, in semifluid fashion, or otherwise modify their mutual relationships, allowing the film itself to adopt a wide variety

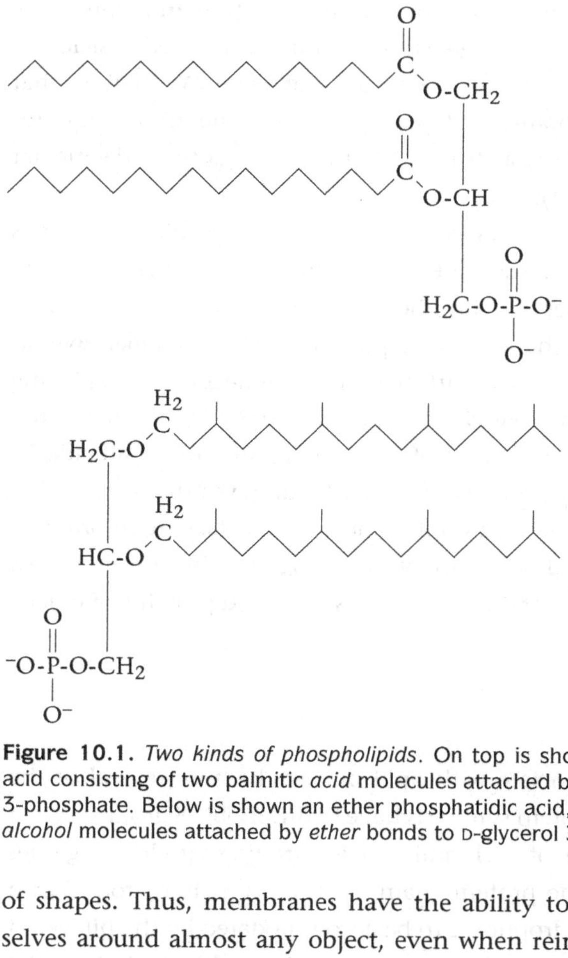

Figure 10.1. *Two kinds of phospholipids*. On top is shown an ester phosphatidic acid consisting of two palmitic *acid* molecules attached by *ester* bonds to L-glycerol 3-phosphate. Below is shown an ether phosphatidic acid, consisting of two phytane *alcohol* molecules attached by *ether* bonds to D-glycerol 3-phosphate.

of shapes. Thus, membranes have the ability to fold and mold themselves around almost any object, even when reinforced by intercalated lipids, such as cholesterol, or by proteins (see below). In some archaeal prokaryotes, especially adapted to very harsh environmental conditions, this flexibility is limited by inter-tail bonds that render the structure more rigid.

Another important property is *self-sealing*. Lipid bilayers and the membranes arising from them have a strong, inherent tendency to form closed sacs, or vesicles. Thanks to the self-sealing property of bilayers, a small needle can be inserted into a cell and withdrawn again without damage, as is done in a number of genetic manipulations. Also, when membrane patches come close enough to each other, they may join and reorganize, leading either to fission of a single structure into two closed vesicles or, alternatively, to fusion of two vesicles into a single one. We shall see that this propensity of membranes for fission and fusion underlies many fundamental cellular processes, such as endocytosis, exocytosis, and vesicular transport (see Chapter 15).

In all biological membranes, the basic phospholipid bilayer fabric is enriched in various ways. The inner, fatty film of the bilayer often provides shelter for a number of hydrophobic molecules. Some of these have a rigid structure and a shape well adapted to establish cooperative attractions of van der Waals type with the bilayer molecules, thus playing an important reinforcing role. Such is the case, notably, of cholesterol, which is an obligatory constituent of the outer membrane of all eukaryotic cells, including ours, and, as such, hardly deserves the opprobrium often too categorically inflicted on it.[1] Cholesterol is lacking in prokaryotes, where it is replaced by a variety of analogues. Most prominently, all natural membranes contain a variety of specialized proteins of crucial importance.

Membrane Proteins

Proteins are able to associate with lipid bilayers because several of the amino acids used for protein synthesis have hydrophobic R groups. When a large enough number of such amino acids are present close together in a given stretch of the protein chain, they form a short, rod-shaped segment sufficiently hydrophobic to be accommodated by the bilayer. If such a segment ends the chain, it serves as anchor, with the bulk of the molecule hanging from one or the other face of the bilayer. If a hydrophobic segment is situated inside a protein chain, it can form a transmembrane bridge, with the two parts of the molecule separated by the segment

---

[1] The importance of cholesterol is illustrated by the fact that any alimentary insufficiency of this substance is automatically compensated in the organism by endogenous synthesis, a process subject to very precise regulation. The fact remains that cholesterol *excess*, especially associated with low density lipoproteins (LDL), the so-called "bad" cholesterol, represents a nonnegligible risk factor of cardiovascular trouble.

protruding on opposite faces of the bilayer. Finally, if, as is often the case, more than one such segments follow each other along the protein chain, the molecule will snake in and out across the bilayer, with both ends coming out on the same face or on opposite faces of the bilayer, depending on whether the number of hydrophobic segments is even or odd.

Membrane proteins have multiple functions. Some are *enzymes*, acting, for example, on hydrophobic substrates, such as fatty acids, which tend to segregate within the lipid bilayers, or on membrane-linked substrates. Thus, the progressive assembly of the oligosaccharide side chains of membrane glycoproteins (see below) is carried out by membrane-linked enzymes. A particularly important group of membrane-embedded catalysts comprises the components of respiratory chains and of photosynthetic centers (see Chapter 11).

Transporters are another important group of membrane proteins. Lipid bilayers are, in spite of their extreme thinness – about 6 nm – highly efficient barriers, impermeable to the majority of water-soluble substances. To cross the inner lipid film of the bilayer, a water-soluble substance must exhibit a sufficient degree of lipophilicity to be able to pass reversibly between water and lipid in appreciable amounts. Small molecules with mildly hydrophilic groups, such as ethanol, the common alcohol, have this property. Most exchanges of matter across biological membranes are mediated by proteins, which either simply facilitate the passage of substances in the thermodynamically favored direction from higher to lower concentration (permeases) or force them in the opposite direction with the help of ATP-derived energy (see Chapter 4). When such active transport systems act on electrically charged substances, they are usually called *pumps*. They create membrane potentials, that is, electric charge disparities between the two areas separated by the membrane. Such disparities, involving sodium (Na^+), potassium (K^+), or chloride (Cl^-) ions, play a key role in nervous conduction. When they involve protons (H^+), they serve to acidify certain sites (lysosomes, gastric juice) and, especially, help generate – or, more frequently, exploit – protonmotive force (see Chapter 12).

Some transporters act as *translocators* of proteins across membranes. It is by such mechanisms that bacteria secrete proteins (exoenzymes) in their surrounding medium and that eukaryotic cells steer proteins synthesized by cytosolic ribosomes to their cellular sites, such as mitochondria, chloroplasts, or elements of the cytomembrane system (see Chapter 15). These mechanisms are directed by special sequences of the protein molecules that serve as labels of their final destination.

Yet another class of membrane proteins is represented by *receptors*, which bear specific groups that have the ability to bind certain defined substances and to respond to this binding in a functionally relevant fashion, such as translocation of the bound substance across the membrane or its ferrying from one site to another by some form of membrane-dependent bulk transport, for example, endocytosis or vesicular transport. A particularly important category of receptors acts in communication. These receptors bind certain substances present on one side of a membrane and, when doing so, undergo a conformational change such that a significant effect is triggered on the other side of the membrane. This is how, for example, cells respond internally to certain external agents, such as hormones, neurotransmitters, or drugs, that are unable to cross membranes. Surface receptors often consist of glycoproteins, which extend their carbohydrate side chains to the outside like miniature, molecular antennae.

The Birth of Membranes

In present-day life, the glycerol 3-phosphate molecule that forms the three-carbon core and phosphate head of the large majority of phospholipids (see Figure 10.1) arises by reduction from the corresponding ketone, dihydroxyacetone phosphate, a glycolytic intermediate. Note that two distinct chiral forms of the molecule can arise in this way, thus accounting for one of the differences between ester and ether lipids:

$$
\begin{array}{ccc}
\mathrm{CH_2OH} & \mathrm{CH_2OH} & \mathrm{CH_2OH} \\
| & | & | \\
\mathrm{HO\text{-}C\text{-}H} \quad \xleftarrow{\;(+2\,\mathrm{H})\;} & \mathrm{C}{=}\mathrm{O} \quad \xrightarrow{\;(+2\,\mathrm{H})\;} & \mathrm{H\text{-}C\text{-}OH} \\
| & | & | \\
\mathrm{CH_2\text{-}O\text{-}\textcircled{P}} & \mathrm{CH_2\text{-}O\text{-}\textcircled{P}} & \mathrm{CH_2\text{-}O\text{-}\textcircled{P}}
\end{array}
\tag{2}
$$

L-glycerol-3-$\textcircled{P}$ Dihydroxyacetone-$\textcircled{P}$ D-glycerol-3-$\textcircled{P}$

(ester lipids)   (glycolysis)   (ether lipids)

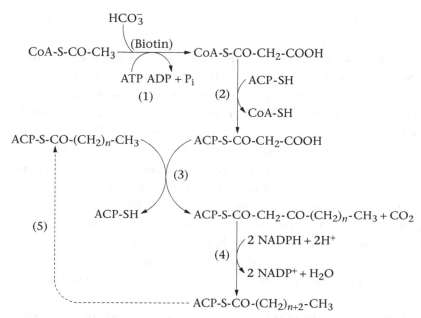

Figure 10.2. *Fatty acid synthesis.* In step 1, acetyl coenzyme A is carboxylated to malonyl coenzyme A with the help of energy provided by the splitting of ATP to ADP and inorganic phosphate. This reaction takes place by sequential group transfer (see the section "Group Transfer..." in Chapter 4), with carboxyl phosphate as double-headed intermediate and biotin serving as carrier of the carboxyl group. In step 2, the malonyl group is transferred from coenzyme A to the thiol group of the phosphopantetheine prosthetic group of acyl-carrier protein (ACP). In step 3, the growing fatty acid chain is transferred from its ACP carrier to the methylene group of malonyl ACP, with release of carbon dioxide and formation of a β-ketoacyl ACP derivative elongated by two carbon atoms. In step 4, this derivative is reduced with NADPH as electron donor, giving rise to the corresponding saturated acyl ACP, ready to participate in a new elongation cycle (step 5). Note that n is an even number. It equals zero at the start of synthesis, the donor in step 3 being acetyl coenzyme A.

Long-chain fatty acids, which make up the tails of ester lipids, are assembled by the successive addition of two-carbon acetyl units, with coenzyme A or pantetheine phosphate (see "Thioesters and Group Transfers" in Chapter 6) as carrier. The mechanism involved is shown schematically in Figure 10.2. It has several peculiarities. First, fatty acids, like proteins (see "Protein Synthesis Today" in Chapter 8), grow by their heads, while always remaining attached to their carrier. This can be seen in step 3 of Figure 10.2, in which the growing chain is transferred to the newly added acetyl unit (rather than the opposite). A second, particularly important peculiarity of the process is that formation of the carbon–carbon bond involved in this transfer requires more energy than is made available by the thioester bond consumed in the reaction. This thermodynamic hurdle is overcome by the prior attachment of a carboxyl group to the receiving methyl end of the acetyl group, which is thereby converted to a malonyl group (step 1). This carboxyl group is released back as carbon dioxide concomitantly with the transfer reaction, so that the energy of decarboxylation is added to the energy of splitting of the thioester bond to support the formation of the carbon–carbon linkage in the growing chain (step 3). Carboxylation (step 1) takes place by a typical sequential group-transfer mechanism, with biotin (vitamin H) as carboxyl carrier. Finally, it will be noted that pantetheine phosphate serves as acyl carrier in two distinct forms, as part of the soluble coenzyme A for transport and as the protein-linked prosthetic group of acyl-carrier protein (ACP) for catalysis.

Coenzyme A also serves as acyl carrier in the synthesis of the ester bonds whereby fatty acids are linked to glycerol 3-phosphate, forming phosphatidic acids (see Figure 10.1). The acyl–coenzyme A derivatives may arise directly by transfer of the newly completed acyl group from the acyl-carrier protein to coenzyme A (see Figure 10.2). More frequently, free fatty acids are attached to coenzyme A by sequential group transfer dependent on AMP transfer from ATP [see reaction (15) in Chapter 6]. Note also that if the phosphoryl group of a phosphatidic acid is replaced by a fatty acyl group, then a triester of glycerol, or triglyceride, is obtained. This is the basic structure of the principal animal and vegetable fats and oils, a major reserve substance in the living world.

The long-chain alcohols that make up the hydrophobic tails of ether lipids (see Figure 10.1) arise by the successive addition of five-carbon units provided by a pyrophosphorylated precursor, isopentenyl pyrophosphate, which may also react as its isomer, dimethylallyl pyrophosphate:

$$CH_3\text{-}\overset{\overset{\displaystyle CH_2}{\|}}{CH}\text{-}CH_2\text{-}CH_2\text{-}O\text{-}\overset{\overset{\displaystyle O}{\|}}{\underset{\underset{\displaystyle O^-}{|}}{P}}\text{-}O\text{-}\overset{\overset{\displaystyle O}{\|}}{\underset{\underset{\displaystyle O^-}{|}}{P}}\text{-}O^- \;\rightleftharpoons\; CH_3\text{-}\overset{\overset{\displaystyle CH_3}{|}}{CH}\text{=}CH\text{-}CH_2\text{-}O\text{-}\overset{\overset{\displaystyle O}{\|}}{\underset{\underset{\displaystyle O^-}{|}}{P}}\text{-}O\text{-}\overset{\overset{\displaystyle O}{\|}}{\underset{\underset{\displaystyle O^-}{|}}{P}}\text{-}O^-$$

  isopentenyl pyrophosphate   dimethylallyl pyrophosphate

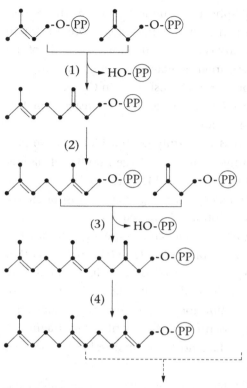

Figure 10.3. *Isoprenoid synthesis.* In step 1, the dimethylallyl group of dimethylallyl pyrophosphate, on the left, is transferred to the isopentenyl pyrophosphate, on the right (see text for complete molecular structures), with release of inorganic pyrophosphate. In step 2, the isopentenyl head of the 10-carbon molecule formed is isomerized to the dimethylallyl configuration, ready to be transferred (step 3) to a new isopentenyl pyrophosphate, with release of inorganic pyrophosphate and formation of a 15-carbon chain. Isomerization of this molecule takes place again in step 4, preparing another elongation round that will give rise to a 20-carbon chain. Double bonds are subsequently reduced to a greater or lesser extent.

As shown in Figure 10.3, the growing chain with a dimethylallyl end is donated, with release of inorganic pyrophosphate, to an isopentenyl pyrophosphate molecule, which subsequently isomerizes to the dimethylallyl configuration before donation of the elongated chain to a new isopentenyl pyrophosphate molecule. This process continues until completion of the chain. The double bonds are later reduced to a greater or lesser extent, following which the long-chain alcohol (for example, the 20-carbon phytanol molecule shown schematically in Figure 10.1) is donated to a hydroxyl group of the glycerol 3-phosphate molecule, again with release of inorganic pyrophosphate. It will be noted that, once more, the chain grows by its head. Also noteworthy is that we encounter here one of the rare cases in which a biosynthetic intermediate is activated in pyrophosphorylated form (see the section "Group Transfer..." in Chapter 4). This form, incidentally, is not produced by pyrophosphoryl transfer from ATP, but by two successive phosphoryl transfers.

The mechanism of Figure 10.3 is of truly central biological importance. It is involved in the synthesis of a very large number of biological molecules in addition to the hydrophobic chains of ether lipids. Among these molecules, which make up the large family of isoprenoids, are phytol, a side chain of the chlorophyll molecule; quinone electron carriers (see Chapter 11); long lipid-soluble chains that serve to anchor certain proteins and other water-soluble molecules; vitamin A and other carotenoids, including visual pigments, as well as other lipid-soluble vitamins (D, E, K); cholesterol and the other derivatives of the sterol nucleus, including corticosteroids and sex hormones; latex, the parent substance of rubber; and the innumerable essential oils that confer their fragrances on aromatic leaves and flowers. In fact, no organism devoid of isoprenoid molecules exists in the entire biosphere.

As to the parent isopentenyl nucleus, it arises, in most organisms, from a key component, known as mevalonic acid, which itself forms from three coenzyme-A-linked acetyl groups.[2] At this level, therefore, fatty acids and isoprenoids share a common origin.

Faced with the existence of two entirely distinct kinds of membrane phospholipids, we are naturally led to ask which came first. This question is directly related to the phylogeny of prokaryotes, because, as mentioned above, Bacteria (and eukaryotes) have ester lipids in their membranes, whereas Archaea have ether lipids. This problem will be discussed in Chapter 14.

---

[2] In certain bacteria and in the chloroplasts of algae and plant cells, the synthesis of the isopentenyl nucleus takes place by a different pathway, involving carbohydrate constituents (Rohmer, 1999; Rohmer et al., 2004). I am indebted to Guy Ourisson for having drawn my attention to this fact.

Looking at the question in the framework of present-day biochemistry, one is first struck by the fact that isoprenoids are much more "interesting" than fatty acids: they include electron carriers, photoreceptors, vitamins, hormones, pigments, scents, and other biologically valuable substances, whereas almost the only virtue of fatty acids is to store energy in compact form. As every diet-conscious person knows, the combustion of fat releases twice as many calories per unit weight as that of carbohydrates or proteins. Second, the biosynthesis of isoprenoids is simpler than that of fatty acids, which, as we have seen, require a complex, dual activation mechanism for their making. It is thus tempting to assume that isoprenoids came first. This is the thesis vigorously defended, largely on the strength of fossil evidence, by the French chemist Guy Ourisson (Ourisson and Nakatani, 1994). Recognizing that actual phospholipids may not have arisen readily under prebiotic conditions owing to the complexity of their polar heads, Ourisson has proposed that the first amphiphilic molecules may have been simple phosphate esters of isoprenoid alcohols; these have been synthesized and have indeed been found to form membranes and vesicles.

The American investigator David Deamer (1998), while accepting Ourisson's thesis with respect to early life, doubts that it is relevant to the prebiotic situation, which, he believes, may not have been conducive to the formation of isoprenoids. He proposes, instead, that the first bilayers may have consisted of fatty acid molecules. Deamer has found that the Murchison meteorite contains amphiphilic substances capable of forming vesicles. One of these substances has been tentatively identified as a nine-carbon fatty acid, which, when prepared synthetically, has been seen to assemble into vesicles under certain conditions. The possibility that the first membranes may have consisted of simple fatty acids has been advocated by other investigators (Hanczyc et al., 2003), who have made the interesting observation that the formation of vesicles from fatty acid micelles is catalyzed by montmorillonite, a clay mineral.

Not all models proposed for the first membranes call on lipid bilayers. The late American biochemist Sidney Fox, one of the pioneers of origin-of-life research, is famous for having, in the 1950s, condensed mixtures of amino acids by dry heat into what he called "proteinoids," which, when exposed to water, gave rise to vesicular structures, or *microspheres* (Fox, 1988). Fox, who died in 1998, devoted the rest of his career to the study of these microspheres, which he believed to be models of the first protocells and for which he made increasingly wild claims of lifelike properties. The resulting discredit that has fallen on his work should not, however, extend to the possibility, which seems plausible, that the earliest membranes could have been formed by peptides. It has also been suggested that the early membranes may have consisted of mineral films, of metal sulfides, for example (Martin and Russell, 2003).

The possibility that my hypothetical multimers may have played a role in early membrane formation also deserves to be considered. If the relative abundance of amino acids at the time life started was anything like that found in meteorites or in Miller's flasks, they would have included many more hydrophobic than hydrophilic molecules. Thus, the multimers derived from these amino acids would have been predominantly hydrophobic. Such molecules could have fitted into primitive bilayers. Or they could, perhaps, even have assembled into membrane-like structures.

All these speculations may be entertained because it is very possible that the first membranes were of simple structure and that phospholipid bilayers replaced the early fabric at a later stage. This possibility fits in with what is known of membrane formation today. In present-day cells, membranes never arise *de novo*; they grow by accretion, that is, by the insertion of new molecules into a preexisting fabric. Thus, membranes come from preexisting membranes, linked by uninterrupted filiation to an ancestral membrane that may go back to the earliest days of life on Earth. This kind of development allows for an evolutionary process in which the original components of the membranes have long disappeared, to be replaced by more elaborate molecules as metabolic capabilities grew, up to today's phospholipids and other complex constituents.

The problem of membrane formation is intimately linked to the appearance of the first cellular structures. The timing of this phenomenon remains the object of debate. Some workers believe that the formation of vesicles was the first step in the development of life. Others are of the opinion that life started in some unstructured "soup" or in the form of a surface-linked monomolecular layer.

Whatever the exact timing of the birth of the first protocells, it must have been early. Even the latest allowable limit antedates the appearance of the first protein enzymes and, probably also, that of many ribozymes. As pointed out in Chapter 3, selection on the strength of the usefulness of a catalytic activity obligatorily requires the existence of a large number of competing protocells. Thus, the ingredients of the first membranes must have been products either of cosmic chemistry or of primitive protometabolism.

A problem attending all early membrane models is the need for selective permeability. A primitive fabric obviously could not have included the kind of special transport systems found in present-day membranes. Therefore, the primitive membrane must have been permeable to entering nutrients and exiting waste products, while being impermeable to larger molecules, in particular the first RNA and protein molecules. The participation of multimers in the fabric of the first membranes could have satisfied this requirement, by providing relatively porous boundaries. Lipids could also have done the job, however. It has been found by Deamer that the required selective permeability can be achieved with phospholipids simply by reducing the length of the hydrophobic tails. In a remarkable experiment, he and his group have built an artificially encapsulated system containing an enzyme that uses ADP to construct poly-A, an all-A, RNA-like substance. With bilayers made of phospholipids containing 14 carbon atoms in their hydrophobic tails, external ADP could enter the vesicles freely, whereas the enzyme and its product remained inside (Chakrabarti et al., 1994).

A similar approach has been taken by the group of Pier Luigi Luisi, in Switzerland (Luisi, 2002). Using vesicles made with oleic acid, a fatty acid, the investigators have constructed a number of fascinating encapsulated systems, including one that catalyzed the polymerase chain reaction (PCR), a system widely used for DNA amplification, which required no fewer than nine components, as well as a high-temperature step.

Another important property of cells that must have been a feature already of the earliest protocells is the ability to grow and multiply by *division.* It is very interesting, in this connection, that even simple, artificial vesicular systems have been found to display growth and division *in vitro* (Szostak et al., 2001; Luisi, 2002; Hanczyc et al., 2003). Thus, the kind of Darwinian competition that has to be postulated to account for early selection phenomena could have taken place with primitive encapsulated systems (see Chapter 3).

11 Protonmotive Force

As explained in Chapters 4 and 5, the ability of living organisms to perform the various kinds of work whereby they subsist and proliferate rests overwhelmingly on the *coupling* between downhill electron transfer and the assembly of ATP from ADP and inorganic phosphate. We have seen in Chapter 6 how this coupling can be accomplished by thioester-dependent, substrate-level phosphorylation mechanisms. Although of immense qualitative importance, these processes account for only a minimal fraction of the ATP produced in most organisms. By far the greater part of the ATP used in the biosphere is assembled by *carrier-level* phosphorylations operating by way of *protonmotive force*, yet another of life's amazing singularities.

Anatomy of a Protonmotive Coupling Engine

Protonmotive machineries are obligatorily housed in the matrix of a membrane impermeable to protons. As shown schematically in Figure 11.1, they consist essentially of two reversible proton pumps oriented in the same direction, one driven by the transfer of electrons between two carriers, and the other by ATP hydrolysis. By forcing protons from one side of the membrane to the other, the pumps create a proton potential that tends to oppose the further translocation of protons. If the

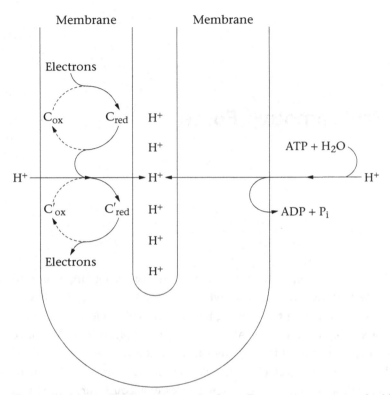

Figure 11.1. *Coupling by protonmotive force.* Two proton pumps, embedded in a proton-tight membrane and oriented in the same direction, drive protons from one side of the membrane to the other, creating an antagonistic proton potential. The pump on the left is activated by the transfer of electrons from the reduced form of one carrier (C_{red}) to the oxidized form of an adjacent carrier (C'_{ox}) of higher redox potential (lower energy level). The pump on the right is activated by ATP hydrolysis. If, as is usually the case, the electron-driven pump is "stronger," that is, builds a higher proton potential than the ATP-driven pump, the latter functions in reverse: electron transfer powers ATP assembly by way of protonmotive force. The opposite takes place if the ATP-driven pump is the stronger: ATP hydrolysis powers reverse electron transfer, that is, from a lower to a higher energy level.

potential built by one pump exceeds the maximum potential attainable by the other pump, it will cause the weaker pump to function in reverse, thus accounting for coupling between the two systems.

As a rule, the metabolically supported, electron-driven pump is stronger than the ATP-driven pump, which is continually weakened by the consumption of ATP for the performance of work: downhill electron flow powers ATP assembly. The opposite may, however, happen under certain circumstances, in particular in chemotrophs, which need to energize electrons for biosynthetic reductions (see below).

Up to three electron-driven proton pumps may be associated in series within the membrane, physically organized in the order of increasing redox potential (decreasing energy level) and linked by intercalated carriers in a manner such that electrons readily flow from one pump to its neighbor. Such electron-transport chains, as they are called, are connected to the outside by entry ports for high-energy electrons and exit ports for energy-depleted electrons (see Figure 11.2).

The most elaborate electron-transport chains are found in certain aerobic bacteria and in the mitochondria of eukaryotic cells, which are descendants from such bacteria (see Chapter 17). These chains contain some fifteen carriers belonging essentially to four distinct families. First are the iron–sulfur proteins, which, as their name indicates, are proteins containing one or more iron ions nestled within clusters of sulfur atoms provided by cysteine residues of the protein and by sulfide ions. Exhibiting a wide variety of redox potentials, these molecules act by the oscillation of their iron ions between the ferrous (Fe^{2+}) and the ferric (Fe^{3+}) state.

A second group of carriers includes flavoproteins, which have as prosthetic group a nucleotide-like derivative of riboflavin, or vitamin B_2, a yellow (*flavus* in Latin), nitrogenous substance that acts as hydrogen carrier. Flavoproteins usually occupy the upper energy level (lower redox potential) of electron-transport chains.

At about the same level is found a lipid-soluble carrier called coenzyme Q, or ubiquinone, which consists of a quinone derivative to which is attached a long isoprenoid chain (see Chapter 10). This molecule alternates between the reduced hydroquinone and the oxidized quinone forms:

$$+ 2\,(-) + 2\,H^+ \qquad (1)$$

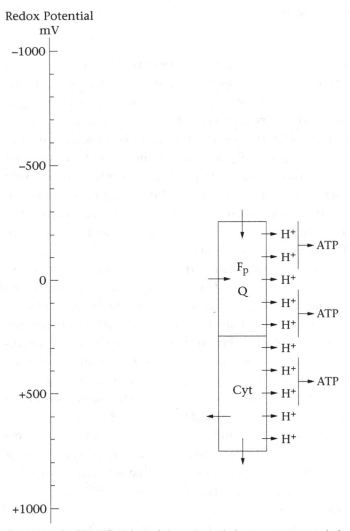

Figure 11.2. *Simplified view of the universal electron-transport chain* (compare with Figure 5.1). The chain consists of up to fifteen membrane-embedded electron carriers organized in order of decreasing energy level (increasing redox potential). The chain is schematically divided into two energy regions: an upper one (lower redox potential), occupied mostly by flavoproteins (Fp) and quinone carriers (Q), and a lower one (higher redox potential), occupied mostly by cytochromes (Cyt), which are carriers of hemoprotein nature. The nature and number of the carriers vary according to the organism concerned but conform, basically, to the scheme shown. Not represented are additional carriers, mainly iron–sulfur proteins and metal ions (copper). Electrons (arrows) enter the chain at the top and leave it at the bottom, generating on their passage a proton potential that, in turn, supports ATP assembly (up to three molecules of ATP per electron pair).

The lower energy level of electron-transport chains is occupied by cytochromes, which belong to the vast group of hemoproteins. These are colored proteins, of hue ranging from pink to brown, sometimes green, that contain a heme as prosthetic group. Hemes are derivatives of the porphyrin nucleus, a flat, polycyclic molecule with a central hole surrounded by four nitrogen atoms. In hemes, this hole is filled by an iron ion, which serves as electron carrier by alternating between the ferrous and ferric states. Other hemoproteins include the oxygen-carrying hemoglobins, in which the iron is permanently in the ferrous state, and a number of hydrogen-peroxide-utilizing peroxidases and catalases, in which the iron remains in the ferric state. Other metals besides iron, copper in particular, often participate in the flow of electrons.

The same kind of organization, but made up of different molecular species and often more or less truncated, is found in a number of aerobic and anaerobic bacteria, as well as in all phototrophic organisms. In the latter, the electron-transport chains are invariably associated with photosystems, which act as light-activated "electron elevators" (see below). The central components of photosystems are the green chlorophyll molecules, which, like hemes, are derivatives of the porphyrin nucleus, but with magnesium, instead of iron, occupying the central hole and with an attached isoprenoid tail serving as lipid-soluble anchor. In photosystems, chlorophylls are usually associated with other pigmented, light-catching molecules, in particular the orange carotenoids, which are relatives of vitamin A and are among the many members of the versatile isoprenoid family.

The ATP-powered proton pump is a remarkable rotatory engine constructed with a number of protein subunits forming a fixed, membrane-embedded "stator" and a central proton-driven "rotor" harnessed to an ATP-assembling catalytic site. This "turbine" functions reversibly, the sense of rotation – and the direction of proton transfer – depending on the magnitude of the proton potential. If, as is generally the case, this potential exceeds the value needed for ATP assembly, the pump makes ATP, while the protons that provide the driving force of this process return to the other side of the membrane, to be repowered by downhill electron flow. In the opposite case, ATP splitting dominates, forcing electrons to travel uphill. We shall see the significance of this phenomenon in chemotrophy.

Metabolic Functions of Protonmotive Electron Transport

In heterotrophic organisms (see Figure 11.3), the electrons fed into electron-transport chains are derived by metabolism from foodstuffs and, during fasting, from reserve substances (such as fat or glycogen) and, if need be, from the organism's own proteins and other active components. Metabolic electrons are mostly funneled into the top of the chains by the soluble electron carrier NADH. A lateral entry port, at the flavoprotein–coenzyme Q level, admits electrons produced by the generation of a double bond, $-CH=CH-$, from a saturated carbon chain, $-CH_2-CH_2-$, situated in the $\alpha-\beta$ position of certain acid derivatives. The dehydrogenation of succinate in the Krebs cycle and that of some acyl–coenzyme A intermediates in the β-oxidation of fatty acids are typical examples of such a reaction.

In most organisms, molecular oxygen (forming water) is the final electron acceptor. Electron-transport chains are often called respiratory chains for this reason. With NADH as donor and oxygen as acceptor, the full capacity of the chains is exploited: three ATP molecules are assembled for each electron pair transferred. In a number of bacteria, oxygen is replaced as final electron acceptor by some mineral, for example, sulfate or nitrate. The ATP yield per electron pair may be correspondingly lowered, depending on the redox potential of the acceptor.

Chemotrophic bacteria (see Figure 11.4) receive all their electrons from mineral donors and most often use oxygen (occasionally some mineral substance) as acceptor. Except for a frequently lower yield of ATP per pair of electrons transferred from donor to acceptor, chemotrophs function very much like heterotrophs, which they in fact mimic when obliged to use their own reserves or substance (not shown in the figure).

A key difference between the two groups of organisms, as already mentioned in Chapter 5, is that chemotrophs have to synthesize all their constituents from simple mineral building blocks and require a large amount of high-energy electrons to this end. Here is where the reversibility of

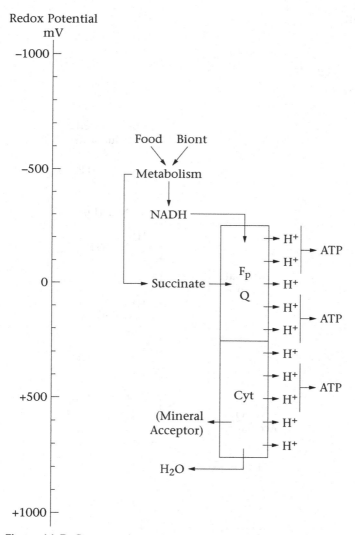

Figure 11.3. *Protonmotive electron flow in heterotrophs* (compare with Figure 5.2). High-energy electrons, provided by foodstuffs or, during fasting, by reserve substances and, if need be, active cell components (Biont), are channeled by metabolism and fed mostly into the top of the electron-transport chain by way of NADH. A small fraction of metabolic electrons enter the chain at the Fp–Q level. The oxidation of succinate to fumarate in the Krebs cycle and the comparable process that occurs in the β-oxidation of fatty acids are reactions of this sort. The main electron exit from the chain leads to molecular oxygen (not shown explicitly), which is reduced to water. In some bacteria, spent electrons are delivered to a mineral acceptor.

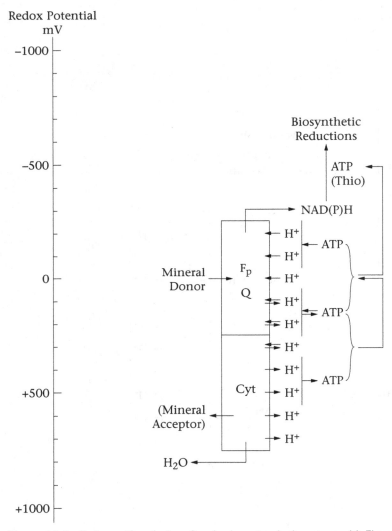

Figure 11.4. *Protonmotive electron flow in chemotrophs* (compare with Figure 5.4). Electrons are provided at an intermediate energy level by a mineral donor. They are most often accepted by oxygen, forming water; occasionally, by a mineral acceptor. Protonmotive force generated in the lower part of the chain supports, directly or by way of ATP, uphill electron transfer in the upper part of the chain, leading to NADH or NADPH, which in turn provides electrons for biosynthetic reductions with the help of ATP produced by the chain and acting by way of thioesters (see Chapter 6).

proton pumps, discussed above, comes into play. The proton potential built in the lower part of the chain serves to reverse, either directly or by way of ATP, the flow of electrons through the upper part, thus allowing the reduction of NAD (or NADP) by electrons provided at a lower energy level by the mineral donor. This phenomenon is helped by the fact that the cytochrome-linked, electron-driven proton pump at the bottom of the chain is considerably more powerful – spans a much greater redox potential difference – than the upper two, especially with oxygen as acceptor. In fact, it cannot be reversed with the help of ATP: aerobic organisms, even amply supplied with energy, are unable to produce molecular oxygen from water. Only phototrophs can do so, as will be seen.

It will be remembered in this connection that electrons provided by NAD(P)H cannot be used as such for biosynthetic reductions; they need to be further energized. This step is also fueled by ATP, but by way of thioesters, as mentioned in the section "Thioesters and Electron Transfer" in Chapter 6.

The situation that obtains in the more primitive phototrophic bacteria is illustrated in Figure 11.5. Light absorbed by photosystem I lifts chlorophyll electrons from ground level to a level some 1500 mV higher in energy (lower in potential), from which the electrons fall down to an iron–sulfur carrier called ferredoxin. Two pathways are open for the electrons exiting ferredoxin. One pathway leads into a truncated, phosphorylating electron-transfer chain, from which the electrons return to chlorophyll at ground level, ready to start the same cycle. Electrons following this pathway thus may cycle endlessly, without the participation of any exogenous donor or acceptor, allowing light energy to power ATP assembly (cyclic photophosphorylation).

A second pathway diverts electrons from ferredoxin into NAD(P)H for the support of biosynthetic reductions, which occur, as in chemotrophs, with the help of a thioester-mediated, energizing step powered by ATP. Electrons used in this manner are replaced by electrons from an exogenous mineral donor. Surprisingly, no system apparently exists whereby photoenergized electrons can be used directly for biosynthetic reductions, even though their energy level would be amply sufficient for this. As shown in Figure 11.5, the electrons fall down unproductively to NAD(P)H,

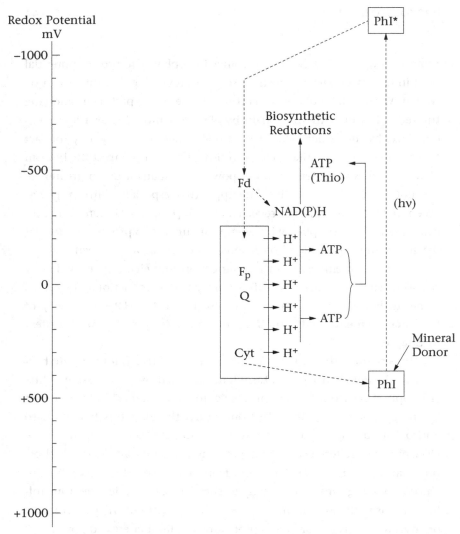

Figure 11.5. *Protonmotive electron flow in primitive phototrophs* (compare with Figure 5.5). The only photosystem of primitive photosynthetic bacteria is photosystem I. In this system, electrons are lifted with the help of light energy (*hν*) from ground level (PhI) to the excited level (PhI*), from which they fall into a protonmotive electron transport chain via a special iron–sulfur protein called ferredoxin (Fd). At their exit from this chain, after supporting ATP assembly, the electrons are returned to the photosystem at ground level, ready to participate in a new cycle (cyclic photophosphorylation). Electrons used for biosynthetic reductions are diverted from ferredoxin to NADH or NADPH, from which they are lifted to the required energy level with the help of ATP produced by the chain and acting by way of thioesters, as in chemotrophs. These electrons are replaced at the expense of a mineral donor, which feeds them into the photosystem at ground level. Note that a substantial part of the energy of the absorbed light is not productively utilized.

142

subsequently to be raised again to a higher energy level with the help of ATP.

Cyanobacteria, together with all eukaryotic phototrophs, which are equipped with chloroplasts derived from ancestral cyanobacteria (see Chapter 17), possess the same photosystem I and associated biosynthetic reductions system as primitive phototrophs; but they do not derive the electrons they use for biosynthetic reductions from some mineral donor; they derive them from *water*, releasing molecular oxygen. This remarkable feat is accomplished by a second photosystem, derived by evolution from photosystem I. As shown in Figure 11.6, photosystem II, when excited by light, lifts electrons over the same 1500-mV span as photosystem I but accepts them lower down, below the energy level (above the redox potential) of the water/oxygen couple, which serves as donor. A complex, manganese-containing enzyme catalyzes this crucial reaction, which solely accounts for the appearance and maintenance of oxygen in the Earth's atmosphere (see Chapter 16).

The water-derived electrons energized by light in photosystem II are fed into the lower part of photosystem I by way of the electron-transport chain, fueling ATP assembly before being reactivated photochemically to serve for biosynthetic reductions (see Figure 11.6). Thus, in the higher phototrophs, all the electrons used for biosynthetic reductions come from water and follow a pathway that includes the support of ATP assembly by what is known as noncyclic photophosphorylation. Any ATP requirement above what is satisfied in this manner takes place by cyclic photophosphorylation dependent on photosystem I.

The composite picture of Figure 11.7 gives a global view of electron flow in the biosphere. The immense importance of the water/oxygen thoroughfare will be noted. It connects aerobic life with oxygenic phototrophy and regulates the level of atmospheric oxygen by the balance between these two processes. As will be mentioned in Chapter 16, this balance has varied in the past and could do so in the future. In addition to these oxygen-centered phenomena, living organisms modify the redox state of their environment by the electron exchanges they bring about in the mineral world. Sulfur, iron, and nitrogen components are particularly affected by these phenomena.

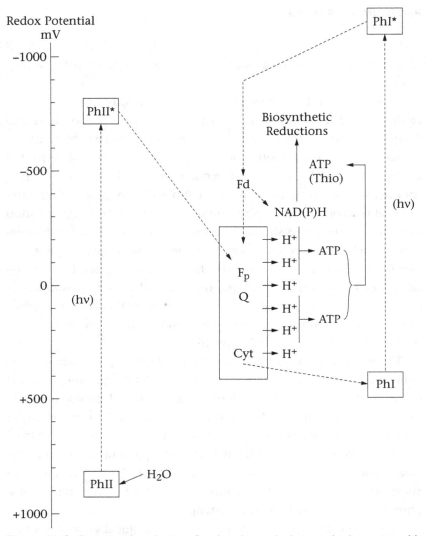

Figure 11.6. *Protonmotive electron flow in advanced phototrophs* (compare with Figure 5.5). Cyanobacteria and chloroplasts, the cyanobacteria-derived photosynthetic organelles of eukaryotic phototrophs (see Chapter 17), possess photosystem I, like their more primitive homologues, and use it in essentially the same way (see Figure 11.5), except for one major difference: electrons used for biosynthetic reductions are not provided by a mineral donor but by water, thanks to a second photosystem. Water electrons are delivered (with the release of molecular oxygen, not shown) into photosystem II at ground level (PhII), with the help of a highly specialized catalytic system involving manganese ions. After activation by light, the electrons are fed from the excited photosystem II (PhII*) into photosystem I at ground level (PhI), by a pathway that traverses the protonmotive electron transport chain and allows the assembly of ATP (noncyclic photophosphorylation). Note that, again, a substantial part of the energy of the absorbed light is not productively utilized.

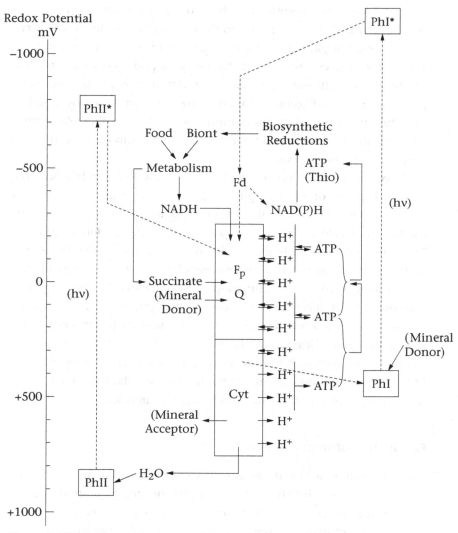

Figure 11.7. *Protonmotive electron flow in the biosphere.* A composite of Figures 11.3 to 11.6, the figure includes a view of how autotrophs deprived of their source of energy can survive by using their own reserves and substance (Biont) in a manner similar to the heterotrophic mode (Figure 11.3).

There are a few exceptions to the above generalizations. It happens occasionally that ATP is bypassed in the exploitation of electron-flow energy. For example, as mentioned earlier, protonmotive force generated in one part of an electron-transport chain may support reverse electron flow in another part without the mediation of ATP. Another remarkable example is the bacterial flagellum, a rotating helical shaft driven by a molecular "turbine" powered by protonmotive force. The possible relationship between this machinery and the rotatory, ATP-assembling proton pump raises an intriguing question.

In very rare cases, electrons are bypassed. This occurs in halobacteria, a special group of phototrophic organisms that thrive in saturated brine and have the unique distinction of not using chlorophyll as light-catching molecule, but rather a purple pigment called bacteriorhodopsin. This substance, which is chemically related to rhodopsin, the visual pigment of the retina, is a protein associated with a carotenoid that acts as light-catcher. Bacteriorhodopsin is embedded in the peripheral membrane of halobacteria. When excited by light, the molecule returns to ground level in a way that uses the energy to drive protons across the membrane, generating protonmotive force, which, in turn, powers ATP assembly. This remarkable photosystem presumably arose independently of the chlorophyll-dependent system. It never supplanted the latter but may have survived in the visual machineries of animals.

Origin of Protonmotive Force

How such elaborate systems as those involved in coupling by protonmotive force ever came into being raises a highly intriguing question. Barring intelligent design, only selection can account for such a remarkable development. Accepting this hypothesis, one is led to ask what circumstances could possibly have favored the appearance of the complex molecular arrangements needed for even the most primitive coupling systems based on protonmotive force.

As suggested in an earlier book (de Duve, 1991), adaptation to an acidic medium appears as an attractive explanation. Acidity is a function of proton concentration (usually expressed by the common logarithm of the reciprocal of this concentration – the pH). The higher the proton concentration, that is, the lower the pH, the stronger the acidity. The inside of cells has a degree of acidity near neutrality, equivalent to a proton concentration of 10^{-7} equivalent per liter (pH 7). Intracellular systems are adapted to this value. Had similarly adapted protocells become exposed to increasing outside acidity (lowering of the external pH), they would presumably have gained a considerable selective advantage from the ability to keep their internal pH constant by expelling protons. Such a process would have required an amount of energy equal to about 2.4 kcal, or 10 kJ, per proton equivalent transferred under physiological conditions, for each order-of-magnitude change in the ratio of outside to inside proton concentration, that is, for each unit difference between the internal and external pH.

Theoretically, such a defense mechanism against increasing acidity could have been acquired in two different ways, depending on whether ATP hydrolysis or downhill electron transfer provided the required energy. We have seen (at the beginning of Chapter 4) that ATP splitting releases about 14 kcal, or 59 kJ, per gram-molecule under physiological conditions. Dividing these values by those given above, we find that an ATP-supported pump could have maintained a constant internal proton concentration against an almost one-million-fold higher outside proton concentration, that is, against an external pH nearly 6 units lower than the internal pH. Protocells possessing such a pump and having available an appropriate, ATP-supplying energy source, would have been able to keep an internal pH of about 7 – physiological for present-day cells – in a medium of pH close to unity, that is, having a proton concentration nearing that of a 0.1 M solution of hydrochloric acid (not far from the limit tolerated by acidophilic prokaryotes, see Chapter 14). The same result could have been accomplished by an electron-driven pump working across a redox potential difference of 600 or 300 mV, depending on whether the electrons traveled singly or in pairs (see the section "Energetics of Electron Transfer" in Chapter 5).

It is tempting to assume that the two kinds of pumps were developed separately, under the control of natural selection, by two distinct protocellular branches threatened by increasing outside acidity. Let the two systems join in a single protocellular population, as a result of lateral gene transfer or of protocellular fusion, and protonmotive force would have been harnessed. One pump could henceforth, under appropriate conditions, force the other to function in reverse. The same result could obviously have been accomplished if the two pumps had arisen jointly in the same protocells. It is, however, more difficult to see how natural selection could have favored the development of both systems side by side, unless their coupling proved beneficial right from the start.

Whether it took place as hypothesized or otherwise, the development of protonmotive systems represents one of the most important singularities in the origin of life. It inaugurated the fundamental carrier-level phosphorylation mechanisms that underlie respiration (including the anaerobic kind that uses electron acceptors other than oxygen) and photosynthesis.

12 Protometabolism Revisited

The account of the earliest beginnings of life offered in the preceding chapters is, perforce, hypothetical. Except for a few scraps of information obtained from extraterrestrial objects and from laboratory simulation experiments, all the steps in my reconstructed history are inferred by reasoning and educated guesswork from what is known of present-day life. With due regard to the weakness of such an approach, it may be profitable to look back over the account and try to discern certain key notions likely to be of more general significance and applicable to any scenario that others may propose.

Overview

Figure 12.1 shows a schematic overview of the early development of life as I have hypothesized it. The diagram depicts three main stages. First there is *abiotic chemistry*, a term used to designate the mechanisms that have generated the raw materials of life. These include, on one hand, the organic building blocks manufactured by cosmic chemistry (meant to include possible terrestrial processes of the kind studied by Miller and his followers) and, on the other, the energy-bearing compounds, pyrophosphates and thioesters, taken to be products of volcanic chemistry (with the help of cosmic chemistry for thioesters).

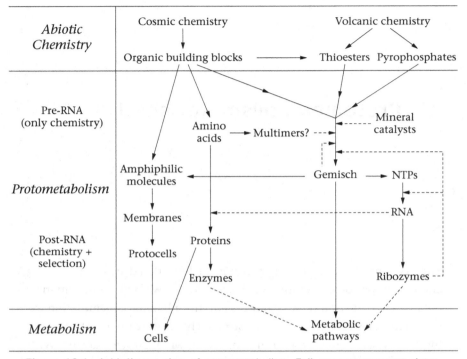

Figure 12.1. *A bird's-eye view of protometabolism.* Full arrows represent chemical processes. Broken arrows represent catalytic influences. For details, see text.

The second stage is *protometabolism*, which is the name given to the set of processes that have led from abiotic chemistry to the third stage, or *metabolism*, defined as the first set of reactions catalyzed by protein enzymes (and, perhaps, ribozymes), prefiguring present-day metabolism and, perhaps, already including certain central systems, such as the glycolytic chain and the Krebs cycle.

The appearance of RNA is represented as a watershed that divides protometabolism into a *pre-RNA* stage, dependent exclusively on chemistry, and a *post-RNA* stage, in which selection is added to chemistry. As imagined, the pre-RNA stage, supported by mineral catalysts and, possibly, by catalytic *multimers* and by protometabolic intermediates with catalytic properties (autocatalysis), characteristically creates a heterogeneous collection of molecules (*gemisch*), among which the four NTP precursors of RNA occupy a position that would undoubtedly not have attracted the attention of an unbiased observer at the time but is revealed to us by hindsight as having been crucial. Conversion of these NTPs to RNA would have initiated replication and, with it, the possibility of selection, ushering in the post-RNA stage of protometabolism (*RNA world*). Concomitantly, or earlier, amphiphilic molecules, produced by cosmic chemistry or by protometabolism or by both, would have assembled into the first membranes, segregating the main protometabolic systems within protocells capable of growth and division.

The most important consequence of the appearance of RNA is the development of protein synthesis, a process that must have gone through a long, complex, evolutionary history. During this period, certain amino acids were recruited; various RNA molecules were progressively selected to become proto-tRNAs, proto-mRNAs, and proto-rRNAs; the genetic code slowly acquired its present structure; and proteins gradually increased in length as the RNA genes themselves became longer thanks to the appearance of more accurate replicating catalysts. As to the RNAs, they have played an essential role in this evolution, notably as catalysts of protein synthesis and, perhaps, of certain RNA reshufflings – perhaps also as ribozymes involved in certain protometabolic or metabolic reactions. Specially noteworthy is the self-perfecting *hypercycle* whereby RNA and replication-catalyzing protein enzymes (or, possibly ribozymes) support each other to allow the formation of RNA and protein molecules of increasing length.

The most problematic part of this overview is the pre-RNA stage of protometabolism. That something of the kind must have happened seems mandatory, but actual mechanisms are totally unknown and wide open to speculation, as even a scant survey of the origin-of-life literature abundantly reveals. The post-RNA stage, by contrast, is perceived with greater confidence on the strength of present knowledge, even though the mechanisms involved remain equally hypothetical and unsupported by direct experimental evidence. With these uncertainties in mind, let us find out what lessons of more general nature can be derived from the proposed scenario.

The Dominance of Chemistry

Given the inescapable premise that there could have been no ribozymes or protein enzymes before the appearance of RNA, the early chemistry must, as a strict minimum, have supplied the means, and the environment must have supplied the conditions, for the construction of AMP, GMP, CMP, and UMP and the activation of those molecules as well as their assembly into polynucleotides.[1] Chemistry and environment must, in addition, have provided the makings of the first protocellular envelopes, which, as we have seen, must have been present at an early stage.

Judging from the results of fifty years of vigorous and inventive research, one gains the impression that cosmic chemistry and prebiotic reactions of the kind investigators have been trying to reproduce experimentally could not, by themselves, have yielded much more than the basic building blocks of the required assemblages. Even at this level, a number of synthetic pathways remain open to question. Most steps toward greater complexity are so far totally unelucidated. My suggested answer to the conundrum, for what it is worth, is summed up in three words: congruence, catalysis, and profligacy.

Congruence, the notion that protometabolism must, in some measure, have prefigured present-day metabolic pathways, rests, as argued in Chapter 3, on the consideration that the early chemistry must have acted as a screen in the selection of the first enzymes and ribozymes, allowing retention of only such catalytic activities as fitted within the chemistry existing at that time. The argument is compelling as regards the earliest selected activities. It is weakened by the possibility that new catalytic activities may have emerged at a later stage and replaced the old ones to a greater or lesser extent. The argument does, however, remain valid for the new activities as well, so that a continuous filiation from protometabolism to metabolism can sensibly be postulated. Certainly, the case is strong enough to justify using the notion of congruence to design experiments (de Duve, 2003).

The view of protometabolism proposed in this book rests on the congruence notion; it postulates, in particular, that RNA was the first

---

[1] I assume here, as the most probable hypothesis, that the first RNAs arose from the same precursors as present-day RNAs. Congruence and parsimony argue in this direction.

replicable information-bearing molecule and that it arose, as it does today, from ATP, GTP, CTP, and UTP, and not in some other way. The NTP
molecules themselves, which obviously cannot have arisen for the "purpose" of making RNA, are seen as products, together with many other
components of a primeval gemisch, of spontaneous chemical events in
which some of their present-day functions, in group transfer, for example, may already have been prefigured. Amphiphilic molecules capable
of assembling into bilayers could have formed as part of this activity. As
to the energy needed for these reactions, it has been assumed, again in
line with the notion of congruence, to rely on pyrophosphates and on
primitive electron transfers, possibly involving thioesters.

Such suggestions would defy credibility were they not linked to the
assumed participation of certain enzyme-like *catalysts*. The need for catalysis in the early chemistry of life is generally recognized. Original to the
present proposal is the suggestion that peptides and other multimers may
have accomplished key catalytic functions in protometabolism, in conjunction with metal elements and, perhaps, with some organic molecules
antecedent to present-day coenzymes. Ease of formation and the fact that
such molecules are most likely to mimic, albeit in rudimentary form, activities carried out by protein molecules are two arguments offered in
support of the multimer hypothesis. It has the advantage of lending itself
to experimental testing (de Duve, 2003).

Finally, *profligacy*, suggested by the term "gemisch," is a notion imposed by the expected lack of specificity of protometabolic reactions.
Simple chemistry and primitive catalysis can hardly have attained together the kind of exquisite specificity manifested by enzyme reactions
today. Hence the likelihood that protometabolism operated in a complex gemisch that was only sorted out later by selection. Enzymes,
thus, were not only selected on the strength of their ability to fit
within protometabolism (congruence); they also served to select, among
protometabolic pathways, those that eventually ended up as part of
metabolism.

It cannot be sufficiently emphasized that all the surmised events
were products exclusively of chemistry, that is, of *deterministic*, reproducible manifestations, entirely dependent on the prevailing physical
and chemical conditions. Selection can have started only after the first

replicable molecules appeared, and its influence must have been gradual, so that protometabolism remained operative for a long time, yielding only slowly and progressively to the kind of self-sufficient and self-contained chemistry carried out by enzyme-catalyzed metabolism.

Looking at the proposed picture of protometabolism with the eyes of a chemist, one must acknowledge that even its minimalist form is complex enough, by virtue of its inescapable requirements – RNA and membranes – to seriously challenge chemical verisimilitude. Defenders of intelligent design have not failed to point this out. Crediting an "unseen hand," however, is no scientific solution. Calling on catalysis seems more rational. In this respect, the multimer hypothesis deserves further investigation (de Duve, 2003), especially in that the multimers themselves are products of a simple chemistry that, as we have seen in Chapter 3, could even take place in outer space.

The Power of Selection

Time and again, in my proposed account of the beginning of life, selection emerges as a decisive factor. Introduced with the very first replicable RNA molecules, this process appears as having shaped every major step of post-RNA protometabolism. Particularly important in this reconstructed picture is the likelihood that, at each step, evolving systems enjoyed the opportunity to explore, extensively if not exhaustively, the range of options offered, with, as end result, *optimization* or near-optimization. This is an extremely powerful situation; it reduces the role of chance to providing enough opportunities so that the possibility most advantageous under the prevailing conditions almost obligatorily emerges. It is the situation described in the introduction under the term "selective bottleneck."

The development of protein synthesis and the optimization of the genetic code are particularly impressive instances of this mechanism at work. So is the progressive appearance of the enzymes that determined the transition from the dirty gemisch of protometabolism to the clean pathways, cleared of side reactions and extraneous junk, that make up metabolism. The stepwise mechanism proposed for protein lengthening provides another telling illustration of the power of selection, as it may have allowed emerging life to reach the infinitesimally minuscule area it occupies in protein sequence space by a succession of optimizing steps. It is noteworthy that it is only at this stage that mopping up of the gemisch could gradually be accomplished. Only when enzyme molecules became progressively longer and thereby gained in specificity and efficacy could certain primitive reactions become replaced by their more efficient, enzyme-catalyzed counterparts, while the others were left to fizzle out. This implies a remarkable degree of robustness and resiliency on the part of protometabolism, which had to sustain incipient life during the many long periods of selection that marked the emergence of metabolism.

The Cradle of Life

Life has left no clue to the site of its birth. It is not even known whether it arose on Earth or elsewhere. One can, however, try to deduce some of the physical–chemical properties of the cradle of life from the presumptive exigencies of protometabolism.

Water was obviously needed, as were the basic organic building blocks most likely provided by cosmic chemistry (and, perhaps, to some extent, by atmospheric chemistry). Huge areas of our young planet (or of some other site) could have fulfilled those requirements. Next, we must ask whether light was needed, which would restrict the cradle of life to surface waters. For reasons that were briefly mentioned before, in particular the intermittent character of sunlight, a photochemical origin of life has been deemed unlikely. Thus, anywhere in water would do. But perhaps not in any waters. It is possible, as suggested by a recent work (Monnard et al., 2004), that the salt concentration played a critical role.

A more searching question is whether life arose within liquid water (the primeval "soup") or on the surface of submerged rocks, as hypothesized by Wächtershäuser (1998). There is no evidence arguing for one or the other possibility. But one must keep in mind the necessary occurrence of early encapsulation. What is needed, therefore, is a site where amphiphilic molecules are likely to assemble into vesicles. Spreading on a surface may have helped such a process (Wächtershäuser, 1998). Vigorous agitation, by waves or jets, is another possibility.

A factor that could drastically limit the number of sites capable of giving birth to life is the requirement, deemed almost inescapable according to our analysis, for pyrophosphates and, presumably also, for hydrogen sulfide in a context conducive to the formation of thiols and thioesters from organic molecules. This condition calls almost obligatorily for a volcanic setting (see Figure 12.1). Such a setting, which has been advocated by others (Washington, 2000), could also provide certain needed metal ions, as well as potential participants in critical electron transfers.

Putting these elements together, one finds volcanic springs and, especially, deep-sea *hydrothermal vents* as most closely fitting the description present-day life seems to reveal to us of its cradle. Discovered in the 1970s, hydrothermal vents are created by cracks in the ocean bottom, spewing back infiltrating water in the form of dark, pressured, superheated, metal-laden, sulfurous jets, or "black smokers." Found to harbor a bizarre collection of microorganisms and even animals, these remarkable formations have been the object of much speculation as potential cradles for life. Special significance has been attached to the possibility that heat could favor the formation of unlikely substances that would subsequently be stabilized by rapid cooling.

This hypothesis has inspired a number of experiments by the Japanese group of Koichiro Matsuno, using a hot–cold flow reactor designed for the purpose of mimicking the alleged conditions. We have seen in Chapter 3 how this group has witnessed the formation of peptides from amino acids in this system. The workers have also observed the oligomerization of AMP, up to the level of trinucleotides (Ogasawara et al., 2000).[2] Intriguingly, the medium contained a small amount of pyrophosphate in the latter experiments. The authors do not explain this choice. Nor do they comment on a possible participation of pyrophosphate in the observed reactions. According to their theory, heat ($110°C$) suffices to provide the energy for the formation of the associations, which are then prevented from breaking down thanks to the rapid cooling. Whether this explanation is thermodynamically plausible requires documentation. The

---

[2] The group has recently described the phosphorylation of AMP to ADP and ATP in a similar system, using trimetaphosphate (a cyclic association of three phosphate molecules joined by pyrophosphate bonds) as phosphoryl donor (Ozawa et al., 2004).

phosphodiester bond between nucleotides is a high-energy bond, equivalent to the pyrophosphate bond. That such a barrier could be overcome by a mere temperature difference of some 100°C would be surprising.

Whatever the conditions, an overwhelmingly important requirement they would have had to meet is to remain *stable* during all the time needed for the continuing support of protometabolism. How long is difficult to estimate. As argued elsewhere (de Duve, 2002), the chemistry itself could have been rapid. But what could have taken a considerable amount of time – possibly as long as several millennia or more – is the long succession of selection events that most likely took place before enzyme-catalyzed metabolism could take over and protocells had become strong enough to withstand environmental disturbances. Volcanic springs or underwater jets could conceivably exhibit the required stability.

The Probability of Life

Life has often been represented as a highly improbable phenomenon, the outcome of an extremely unlikely succession of chance events, a manifestation almost certainly unique in the whole universe, that could very well, but for this extraordinary stroke of luck, never have arisen anywhere, any time. This view, which rests on little more than a quantified sense of wonderment, has, in turn, inspired a variety of philosophical appreciations of the human condition, ranging, remarkably, from utter meaninglessness to supreme cosmic importance, depending on the significance accorded to uniqueness.

The belief that life is unique is no longer widely held by scientists today.[3] Witness the enormous interest in the new discipline, variously known as exobiology, astrobiology, or bioastronomy, that seeks to find evidence of extraterrestrial life. Even extraterrestrial intelligence has been deemed sufficiently likely to warrant costly detection efforts. So far, however, the only evidence alleged in support of extraterrestrial life

---

[3] An exception is the British paleontologist Simon Conway Morris, who, in a recent book (Conway Morris, 2003), has defended the strange thesis that life is probably unique in the entire universe but that, once present, it was bound to give rise to humankind, thus making humans "inevitable in a lonely universe," a philosophically loaded conclusion.

has been the presence of organic compounds, which we have seen are almost certainly products of cosmic rather than of biological chemistry (see Chapter 1).

In discussions of this problem, it is important to distinguish between the probability of the emergence of life under the physical–chemical conditions that prevailed at the site of its birth and the probability of those conditions themselves. With respect to the former, the main conclusion that arises from the proposed analysis is conveyed in the capsule phrase "hardly any chance, mostly necessity." Deterministic chemistry and selective optimization clearly dominate the picture we have arrived at, thus narrowly tying the outcome to the conditions that ruled the chemistry and determined the selection.

Admittedly, one cannot rule out the possibility that some rare substance or some improbable chance occurrence played a decisive, nonreproducible role in the appearance of life. Scientific caution commands such a reservation, but common sense argues against it. One doesn't readily see how a complex network of interconnected chemical reactions, such as must have composed protometabolism, could have been critically influenced by a single, improbable chance factor.

Accepting as a working hypothesis, possibly to be qualified in the light of new findings, that life as we know it was bound to arise under the conditions that obtained at the site of its birth, we are led to conclude that the probability of life is approximately equal to the probability of the physical–chemical conditions under which it arose. If our reconstruction of the cradle of life is correct, this means that the probability of extraterrestrial life depends on there being elsewhere in the universe celestial objects likely to harbor hydrothermal vents physically and chemically similar to those found on Earth.

This question is obviously unanswerable in the present state of our knowledge. In any case, it is not the biologist's business. All that can be said is that the present direction of astronomical research favors multiplicity rather than uniqueness. It is already known that planetary systems are far from rare. To be sure, only large planets orbiting close to their sun have been detected so far, but this is due to technical limitations that may well be overcome by the advances of tomorrow. The existence of Earthlike planets is not excluded; it is seen as likely by many experts. On the whole, sheer numbers would seem to argue against rarity. Considering the number of sunlike stars believed to exist in our Galaxy (on the order of thirty billion) and the estimated number of galaxies in the universe (about one hundred billion), the chances that many Earthlike planets exist would seem to be very high.

There remains the question as to how closely Earthlike a planet has to be to generate life. From what we have seen, liquid water separated from a molten core by a fissured crust may be a prerequisite. Such features may not be uncommon, but do they suffice? There is always the possibility that some special condition, not included in our assessment, has to be fulfilled. Perhaps the magnetic field has to be just right. Or the planet's orbit must have the right degree of ellipticity, its rotation axis the right tilt. Or there may be a need for a moon of just the right mass circling at just the right distance to produce the right tidal movements. One can always, if motivated to do so, multiply parameters to the point of conferring on planet Earth a unique character that makes it the only possible cradle of life (Conway Morris, 2003).

Evidently, until clear evidence of extraterrestrial life is obtained, the matter must remain open. Even then, kinship with Earth life will have to be ruled out. Discovering life on Mars, for example, might not be decisive in itself. It is not considered impossible at present that Martian life could originate from Earth, or terrestrial life from Mars, or both from some third site in the solar system. The two forms of life would have to differ in a significant way for their independent origin to be incontrovertibly established. But is such a difference likely?

The Singularity of Life?

Notice the question mark. It is there because the answer to the question posed could be very different depending on the significance attached to the word "singularity." If it is meant to imply that our planet is the only life-bearing object in the universe, the answer, as we have just seen, is likely to be no: life is probably widespread in the universe. At least, this probability is deemed sufficient by a large segment of the astronomical community to justify a vigorous search for extraterrestrial life. On the other hand, if singularity is taken to apply to the basic chemistry that has been the main object of our discussions, the answer could be affirmative. There may well be no life in the universe other than life as we know it, its chemistry being written in cosmic physics and chemistry.

The deterministic view suggested by our analysis supports such a possibility, as it encompasses most of the key chemical features of life, including membranes, bioenergetic couplings (at least of the substrate-level kind), DNA genes, RNA mediators of gene expression, information transfers via base pairing, RNA-dependent protein synthesis, and the universal genetic code, as well as a number of enzymes, coenzymes, and ribozymes. To the extent that those features arose by a combination of deterministic chemistry and selective optimization, they were essentially bound to arise under the conditions that surrounded their birth and would similarly arise should these conditions be duplicated. Almost the only key feature that may have been decided by chance is chirality, although, even here, there may have been some bias (see Chapter 2).

Such a parochial attitude contrasts with the frequently evoked possibility of there being other life forms, differing from those we know by the nature of their organic constituents and/or by the mechanisms whereby they build their own substance, exploit environmental energy sources, or store, transfer, and express information, perhaps even constructed from elements different from the standard CHNOPS formula. Such open-mindedness is commendable from a purely abstract point of view but is of little concrete value. For simple heuristic reasons, it is preferable to limit our definition of life to what is known and not to indulge in fatuous speculations unlikely to inspire any fruitful experimental approaches.[4] Furthermore, the messages of cosmic chemistry should not go unheeded. When looking at a list of the compounds detected in outer space or in meteorites and other celestial objects, one is struck by the number of substances that are utilized by life. This can hardly be a meaningless coincidence.

---

[4] I do not wish to decry here the kind of research that endeavors to create artificial systems mimicking, perhaps even more efficiently, some key property of living organisms. Such investigations are often fascinating and may lend themselves to valuable applications. Until the created systems are shown to have a relationship to reality, however, they are better called "biomimetic" or "lifelike" than advertised under the misleading name of "artificial life."

13 The LUCA

There is overwhelming proof that all known living organisms descend from a single ancestral form, the LUCA, or *last universal common ancestor*, also called cenancestor by some. Conclusively evidenced by the many singularities already discussed, the single origin of all known forms of life is further demonstrated by the many sequence similarities that exist among genes that perform the same functions in very different organisms, be they microbes, plants, fungi, or animals, including humans. It is clear that such similarities can be explained only on the basis of direct kinship among the genes concerned, leading to the obvious conclusion that the owners of the genes must be similarly related.[1]

This seemingly unassailable conclusion has come under challenge in recent years, on the ground that genes do not only travel vertically, from one generation to another, but may also be exchanged horizontally between two totally unrelated cells. This point is relevant to the construction of molecular phylogenies (see next chapter) but does not seriously undermine the theory of single origin, which rests on a solid body of convergent evidence and allows a number of straightforward considerations, irrespective of phylogenetic details. These considerations will be

---

[1] This fundamental point has been illustrated by a quantitative example in the introduction to de Duve (2002).

the topic of the present chapter, leaving phylogenies to be discussed in the next one.[2]

A Reconstructed Portrait of the LUCA

According to simple logic, it should be easy to arrive at a minimal representation of the LUCA by putting together all the properties shared by all present-day organisms, the rationale being that these properties must all have been inherited from the LUCA. Such a picture will be minimal in that it will not include properties that existed in the LUCA but were later lost in one or more lineages. On the other hand, it may include properties that were absent in the LUCA but were acquired later, either by convergent evolution or through some isolated innovation subsequently spread by horizontal gene transfer.

It must be noted that the risk of omissions is greater than the opposite. Evolutionary losses are commonplace, whereas universally shared evolutionary gains can only have been made very early after LUCA started branching out. Acquisition of the same biochemical property by convergent evolution, already unlikely in the case of two distinct lines, becomes close to impossible for a larger number. As to horizontal gene transfer, one must keep in mind that genes do not travel far on their own. An additional gene, whether born in one of the branches sprouted by ramification from the LUCA or present already in some lineage that existed in LUCA days but later became extinct, could have become part of a common heritage only if it had the opportunity to reach all the other branches that have left surviving lineages. This could have been possible only if the members of those branches lived close enough together to be capable of exchanging genes. Depending on the extent of such gene-sharing, our picture of the LUCA may be more or less fuzzy (see next chapter), but its basic characteristics are likely to be the same.

Today, all living organisms are made of one or more cells, surrounded by membranes constructed with phospholipid bilayers. All utilize DNA as

---

[2] In writing this chapter, I have adopted the common sense view of the LUCA, which is probably shared by a majority of workers in the field. As will be seen in later chapters, some dissenting opinions have been expressed on the topic.

replicable repository of genetic information, express this information by transcription of the DNA into RNA, and translate the RNA into proteins by the same mechanisms, including, with rare exceptions of relatively recent origin, the same genetic code. All rely on the same base-pairing phenomena for their information transfers. All carry out the same general kind of intricate metabolic reactions, built on a common core and catalyzed by protein enzymes of large size, made mostly of several hundred amino acids, with the help of coenzymes of which several are universally distributed. All use ATP as the main energy vehicle and regenerate this substance by coupled, downhill electron transfers, ensured largely by protonmotive, carrier-level phosphorylations, with a most often small but highly significant contribution by thioester-linked, substrate-level phosphorylations.

Taking these common properties as having all been inherited from the LUCA, we arrive at the conclusion that this ancestral organism must have had an essentially "modern" character. Presumably, it was single-celled and resembled, in its general organization, a simpler prokaryote more closely than the more complex eukaryotes.[3] In short, should we encounter the LUCA today, we might well not recognize its extreme antiquity, mistaking it for some primitive, present-day microbe.

This picture leaves one major uncertainty: Was the LUCA autotrophic or heterotrophic? In terms of survival ability, heterotrophy would seem to be the more general property, in that autotrophs share with heterotrophs the capacity to subsist on the breakdown of organic molecules (at least their own reserves and substance). On the other hand, it does not seem very probable that life could still have been supported by products of cosmic chemistry after the long evolutionary pathway that led to the LUCA. It is thus more likely that the LUCA was autotrophic and, if simplicity is to be a criterion, chemotrophic rather than phototrophic.

---

[3] This opinion is not unanimously shared. A small number of investigators defend the thesis that the LUCA had a eukaryotic type of organization and that prokaryotes are descended from this ancestral form by "reductive evolution." I am indebted to Nicolas Glansdorff for providing me with pertinent information on this question (Glansdorff, 2000; Xu and Glansdorff, 2002). As will be pointed out in subsequent chapters, the proposed theory leaves entirely open the problem of the birth of eukaryotic cells. There can be no reduction without prior "complexification."

Another uncertainty is whether the LUCA possessed split genes (see section "The Growth of Proteins" in Chapter 8). This question, which was much debated in the years that followed the discovery of split genes (for the early history of the field, see: de Duve, 1991), is raised by the fact that introns are virtually absent in prokaryotic genes and that their frequency in eukaryotic genes tends to increase with the complexity of the organisms concerned. At first sight, this fact would seem to argue in favor of an evolutionary gain. Nevertheless, the opposed theory of the "antiquity of introns" (Gilbert et al., 1986) has been defended by a number of workers, who believe that prokaryotes lost split genes in the course of a genome "streamlining" process, selectively promoted by the advantages of rapid multiplication.

In addressing this question, a distinction must be made between the *number* of introns and the *ability to splice* mRNA molecules. This ability is necessary even for a single intron and must, therefore, have been present in the first eukaryotes, which could have inherited it from the LUCA, rather than developed it *de novo*. As we have seen, RNA splicing involves ribozymes, which could possibly go back to the early days of genome lengthening, long before the advent of the LUCA. Furthermore, as argued by Poole et al. (1999), the development of new ribozymes in cells amply supplied with sophisticated protein enzymes is unlikely, thus strengthening the view that the LUCA possessed the ability to splice RNA molecules and that prokaryotes lost this ability. This point has been raised in support of the theory that the LUCA may have had a eukaryotic organization (see footnote 3, above). Note, however, that splicing ability has no obvious link with the eukaryotic phenotype (see Figure 15.1). An RNA-splicing LUCA could have a typically prokaryotic phenotype.

Birth of the LUCA

Between the hypothetical history depicted in the preceding chapters and the LUCA, as just reconstructed, lies a long period of maturation about which little more can be said than that it must have happened. In the LUCA, RNA has definitely yielded to DNA as replicable repository of genetic information, while protometabolism has given way to metabolism as the means of chemical support. Genes and the corresponding proteins have completed their long process of progressive lengthening and reached a size that has essentially not changed since. A number of genes are probably subject already to transcriptional regulation by special proteins. Membranes contain basic elements of their present-day attributes, including a minimum number of transporters, pumps, and receptors, together with some mechanism of protonmotive coupling between electron transfer and ATP assembly. Quite possibly, membranes already possess systems for the export of proteins and other materials, allowing the construction of an extracellular wall. For all we know, the cells may even propel themselves with a flagellum powered by protonmotive force. Finally, cell growth is already linked to coordinated gene replication followed by cell division.

It would be futile to even attempt divining how all these developments took place, because there is virtually no information on which to base such conjectures. All we can do is to assume, as reasonable hypotheses, that the events were slow and progressive, proceeding over many millennia, if not millions of years, and that they were dominated by natural selection acting on fortuitous genetic changes. If such was the case, one intriguing question we may be entitled to ask is what accounted for the LUCA's remarkable singularity.

Strict confinement within an environment uniquely compatible with survival provides a possible answer to this question, but it does not appear likely. Organisms of the LUCA's degree of sophistication would be expected to display sufficient flexibility and resiliency to branch into different milieus. Supporting this hypothesis is the discovery of so-called paralogous genes of very ancient origin. In opposition to orthologous genes, which are derived from a single ancestral gene that has undergone different changes in different lineages, paralogous genes originate from two identical gene copies produced in the same organism and left to evolve independently after the initial duplication from which they arose took place. As mentioned in Chapter 8, gene duplication is a very powerful means of genetic "experimentation"; it allows all kinds of variants of a given gene to be put to the test of natural selection while the unmutated copy of the gene continues to fulfill its function. Several paralogous genes have been traced back to duplication events that happened before the advent of the LUCA, suggesting that the precursors of this organism were already engaged in complex evolutionary competitions.

If we take as more probable the hypothesis that the road to the LUCA is one among a number of branches that spread out in those early days, the question arises whether the LUCA was the outcome of a true selective bottleneck (mechanism 2) or of the kind of singularity described in the Introduction under the term of pseudo-bottleneck (mechanism 4), attained, more by chance than by virtue of some selective superiority, through progressive attrition of all the other branches. We have no answer to this question. But we can, assuming the singularity to be due to a selective bottleneck, ask what environmental upheaval could have constrained evolution to let only a single branch pass through, and what particular quality could have given the LUCA its privileged position. Three possible answers to this question come to mind.

A first possibility is that the supply of building blocks produced by cosmic chemistry and abiotic, terrestrial chemistry was running out and that survival depended on the inauguration of autotrophy. If, as has been assumed in this book, life started heterotrophically, a point on which there exists some disagreement (Wächtershäuser, 1998; Morowitz, 1999), a stage must obviously have been reached when the celestial manna became scarce and only organisms capable of subsisting on inorganic raw materials could survive. The LUCA could have squeezed through such a bottleneck by developing the necessary systems in time, presumably by some form of chemotrophy. As mentioned above, emerging life most likely did go through such a bottleneck; but whether it did so as late as assumed seems questionable. The possibility that the abiotic supply of organic building blocks could have covered all the time, perhaps to be counted in millions of years, needed for emerging life to produce the LUCA does not seem very likely.

A second possibility is that progressive acidification of the environment acted as a bottleneck, allowing survival of only those cells that had developed the protonmotive mechanisms needed to protect their internal pH from the acid threat. We have seen that the need to adapt to increasing acidity offers a plausible explanation for the harnessing of protonmotive power (see Chapter 11).

Finally, it is also conceivable that increasing heat, possibly provoked by a large meteorite impact (Sleep et al., 1989; Gogarten-Boekels et al., 1995), was responsible for the bottleneck. This theory is supported by some phylogenetic analyses, to be mentioned in the next chapter, indicating that all the most ancient microorganisms are thermophilic, that is, adapted to high temperatures. However, as we shall see, the significance of these results is being disputed, and not all workers accept the notion that the LUCA was thermophilic.

Note that the possibilities evoked are not exclusive of one another. More than one bottleneck, created by successive environmental crises, could have marked the pathway to the LUCA – for example, in this order or in a different one: starvation, acidification, and excessive heat. Unfortunately, the events we are trying to reconstruct have left no trace other than the time-blurred legacies preserved in present-day genomes. As will be mentioned in the next chapter, the comparative study of gene sequences is an extraordinarily powerful tool for probing the past history of living organisms. But the technology is complex and beset by many difficulties and artifacts on which opinions are far from unanimous.

Viruses

This chapter cannot end without a brief reference to viruses, small entities capable of entering certain cells and proliferating inside them. First recognized by their ability to cause diseases, viruses consist of two main components: a surrounding envelope, or capsid, usually of protein but sometimes more complex; and a small genome coding for the viral proteins. These parts are occasionally accompanied by additional components (mostly enzymes) needed for reproduction of the virus. The viral envelope has the ability to interact with the membrane of susceptible cells in such a way that the viral contents are introduced into the cell. There, the viral genes are replicated, and viral proteins synthesized, by the cell's machineries – eventually with the help of some virus-specific enzyme(s) – with as consequence the formation of new viral particles and, frequently, the death of the invaded cell.

Viruses may be viewed as the ultimate parasites, in that they borrow everything they need for their multiplication from their host cells, carrying as sole baggage what the host cells cannot provide. Virtually all known cells can be attacked by viruses. When bacteria are the target, the viruses are called phages, short for bacteriophages (literally, bacterium eaters). Viral genomes can consist of DNA or RNA, either double-stranded or single-stranded. The RNA viruses themselves fall into two categories depending on their mode of replication. In one group, of which the polio virus is an example, the viral RNA is replicated directly by a specific enzyme (replicase) encoded by the viral RNA. In the other group, exemplified by HIV (human immunodeficiency virus), the causal agent of AIDS (acquired immunodeficiency syndrome), the RNA is reverse-transcribed into DNA by a viral enzyme (reverse transcriptase). This DNA is inserted into the cell genome and transcribed back into RNA for replication and expression. Such viruses are called retroviruses.

Are viruses alive, or not? This question has stirred more philosophical discussions than it deserves. The answer is obviously no, because viruses cannot multiply without the help of a living cell. They are no more alive than is a disk able to play music. A more interesting question concerns the origin of viruses. There was a time when viruses were believed to be intermediates in the long process that led to the first living cells. That view is no longer held.[4] Without a supporting metabolism, viruses can be nothing but inert particles, whether four billion years ago or today. The most probable theory is that viruses are a form of "wandering genes," originating from preexisting living cells, a phenomenon that can happen by a variety of mechanisms of which some are beginning to be understood.

In the case of DNA viruses, such an origin could have taken place at any time, even in the recent past. Things are different for RNA viruses, which have no counterpart in present-day living cells. According to an

---

[4] The recent literature contains occasional references to viruses as having "predated cellular life" (Balter, 2000; Prangishvili, 2003), but without clear proposals concerning the manner in which viruses could have multiplied in the absence of cells. Note that the possibility of entities as complex as viruses arising and thriving in some kind of unstructured, primitive broth that provided the means for their replication is not compatible with the conclusion, arrived at in Chapter 3, that cellularization must have occurred, at the latest, when translation and protein synthesis began to be developed.

attractive hypothesis, RNA viruses may go back to those early protocells that still had RNA genomes or to those that were beginning to convert their genetic information from RNA into DNA (retroviruses).

If this hypothesis is correct, it has interesting implications concerning the stage at which DNA replaced RNA as replicable repository of genetic information. Some viral RNA genomes are of considerable length (up to some 35 000 nucleotides), and the proteins they code for can be of a size similar to that of present-day proteins. The inference, therefore, is that DNA came on the scene at a late stage and that much of the process of gene lengthening and maturation involved RNA, rather than DNA, genes (see Chapters 8 and 9).

14 The First Fork

By definition, the LUCA, being the last universal common ancestor, stands at the apex of the first bifurcation in the history of life that had both branches surviving and extending until the present day through an uninterrupted line of descent. By analogy with other evolutionary bifurcations, it may be assumed that this historical first was initiated when some mutant form was allowed by natural selection to develop into a new line, most likely adapted to a different environment, while the original line went on unchanged. Thus, one branch of the postulated bifurcation was new, whereas the other continued the LUCA, forming what is known as the *root* of the tree of life.[1]

Until the development of modern molecular methodologies, these events remained shrouded in the darkness of the distant past. Our ignorance, in fact, covered the major part of the history of life, when only microbes existed and spread their networks over the entire Earth, leaving

---

[1] The notion of a tree of life, which goes back to Darwin, has come under increasing criticism in recent years. As will be mentioned later in this chapter, a movement is under way tending to replace the tree image by that of a net (reticulated tree) or of a ring. Such representations should not obscure the fact that life stems from a single root, which, by definition, must have split into the first fork, as described here. It is subsequent interrelationships among branches that are responsible for the formation of the proposed net or ring.

hardly any trace of their passage. For most observers, the deepest cleft was that between prokaryotes and eukaryotes (from the Greek *karyon*, nucleus). Prokaryotes, commonly called bacteria until recently, are single cells of small size, with little internal structure and no defined nucleus. Eukaryotes, which may be unicellular (protists) or multicellular (plants, fungi, and animals), consist of much larger cells with a highly structured interior (cytoplasm) and a distinct, fenced-off nucleus. It was generally assumed, almost as a matter of course, that prokaryotes were alone on the scene for some undefined length of time, until a mysterious transition process opened the way toward eukaryotes.

Molecular Phylogenies

This state of our knowledge changed dramatically with the development of methods for the sequencing of proteins and nucleic acids. These methods have allowed a novel, particularly powerful approach to the problem of origins. It was already mentioned in the preceding chapter that sequence similarities among genes that perform the same functions in widely different organisms provide the most convincing proof of the singularity of life on Earth. What are now exploited, with enormous success, for the reconstruction of phylogenetic trees are the sequence *differences* among such genes or their products.

The principle of the new methods is simple. When two lines diverge from a common ancestor, their genes start distinct evolutionary histories. As long as the mutations they undergo remain compatible with survival – we have seen that such is often the case, thanks, notably, to the optimized structure of the genetic code – these mutations are conserved and accumulate in the genomes, which thus become progressively different from that of the common ancestor. Due to their fortuitous character, these mutations are very likely not to be the same in the two lineages. There will thus be sequence differences between the evolutionary offshoots of the same ancestral gene that are found today in the descendants of the two lineages concerned. These differences tell us something of the history of the two lineages since their separation from their last common ancestor.

Up to here, all is luminously simple. Where things start to become complicated is when we try to clarify what is meant by the word
"something."

In an ideal world, mutations would take place with a constant frequency, identical in the two lineages. The number of differences between
the two sequences would thus become a measure of the time that had
elapsed since the bifurcation, expressed in terms of the mutation rate.
This has been called the "molecular clock." Alas! things are not so simple.
They would not be even in an ideal world. Indeed, the same mutation
will, from time to time, take place in the two lineages and will thus go
unnoticed. Or again, a given mutation may be followed by a second one
that supplants or erases it, a happening whose probability increases with
the age of the lineage's ancientness. Especially, not all mutations are point
mutations, leading to the replacement of one base by another. Genomes
may undergo many other modifications, such as deletions or insertions
of one or more bases, inversions and transpositions of certain stretches,
and so on. In addition, it has become clear with experience that the very
notion of molecular clock is false. The frequency of mutations is not constant. It varies from one time or place to another, from one organism to
another, from one gene to another, and even from one region of the same
gene to another.

Because of all this, and much more that I must skip, the construction
of molecular phylogenies has become a very complicated science, beset
by many hazards. To be sure, technologies have been refined accordingly,
increasingly sophisticated algorithms have been invented for computer
processing of the data, and, especially, there is available an invaluable asset: there are a great many genes, and sequencing techniques have been
improved to such an extent that the number of sequences at hand grows
almost daily. Thus, it has become increasingly possible to verify certain
conclusions by cross-checking. This, however, does not prevent even experts from feeling lost today and failing to agree on the relative merits of
different methods.

A detailed discussion of these methods would be far beyond the scope
of this work, let alone the competence of the author. Interested readers

will find a useful introduction to the topic and its abundant literature in a recent in-depth review (Gribaldo and Philippe, 2002). For the purpose of the present book, I can only attempt, with all due caution and awareness of my own limitations,[2] to summarize a few notions that seem to me to emerge with a certain degree of confidence from the moving context of present-day data.

The first attempts at comparative sequencing and phylogenetic analysis date back to a time, in the early 1960s, when only protein sequencing was technically possible (Dickerson and Geis, 1969). First carried out on small proteins of animal and human origin, including cytochrome c, globin, fibrinopeptides, and lysozyme, the investigations immediately yielded gratifying results. Compared with paleontological data, the findings showed excellent agreement, supporting the general validity of the method. They revealed further that the number of amino acid replacements (already corrected for multiple mutations of the same site) was a linear function of geological time, but of a slope that could vary considerably according to the nature of the protein. The notion of a molecular clock was thus validated, but with the important qualification that there is no absolute clock; each protein has its own. Although not unexpected, this finding was greeted with great satisfaction, because of its Darwinian implication. The rate of change was obviously different for different proteins, not so much because they mutated at different rates, as because natural selection allowed a greater or lesser proportion of mutant molecules to pass through, depending on the stringency of the constraints linking the proteins' functions to their amino-acid sequences. This dependence and, with it, the pace of the molecular clock could change if the protein was being adapted to a different function – if, for instance, lysozyme became α-lactalbumin. All this augured very favorably for the future of the method, destined, as already foreseen at the time, not only to reinforce paleontological data but, especially, to extend them to events of which

---

[2] I am greatly indebted to Andrew Roger for an incisive criticism of an early version of this and the subsequent few chapters. He has helped me avoid several mistakes but bears no responsibility for those that remain. My friend and colleague Miklos Müller has also been very helpful.

no fossil vestige exists.[3] Relying entirely on present-day organisms, the molecular approach allowed the past to be read in the traces it has left in extant genomes, back, at least in theory, to the very first events in the history of life.

This approach, reserved until then to a minority of patient and motivated experts, received an enormous boost in the late 1970s, when methods for DNA and RNA sequencing became available. The subsequent advances in the efficiency and rapidity of those methods have spawned a profusion of new results, together with, as already mentioned, a host of theoretical refinements – and complications.

An early landmark in this new stage occurred when the American microbiologist Carl Woese, a pioneering investigator in the field, undertook to apply comparative sequencing to a wide range of samples of the RNA component of the small ribosomal subunit (SSU rRNA), part of the ubiquitous protein-synthesizing machinery (Woese and Fox, 1977; Woese, 1987). His findings yielded two totally unexpected conclusions. One was that there are two, not one, prokaryotic groups that go back to very early times, close to the LUCA itself. Woese called these two groups "archaebacteria" and "eubacteria." His second discovery, perhaps even more startling, was that the eukaryotic line appeared to be equally ancient and to be itself also initiated in LUCA days. Woese later renamed the three groups "Archaea," "Bacteria," and "Eucarya" and proposed to classify them as *domains*, a term meant to stand higher in the hierarchy than even that of *kingdom* (Woese, 2000). A schematic view of the tree based on SSU rRNA sequences is shown in Figure 14.1.

The tree has no timescale. Dating the first fork must rely on the vestiges of life preserved in geological terrains of known age. Because early life was perforce simple and because most geological formations have undergone considerable upheavals, such vestiges are increasingly scarce and difficult to interpret as their antiquity increases. Until recently, the

---

[3] Pioneered by the late Harvard paleontologist Elso Barghoorn, the search for microbial vestiges (microfossils) in very ancient terrains started at about the same time as the molecular approach, yielding a rich crop of findings, with its inevitable accompaniment of interpretative uncertainties and disputes (Schopf, 1999; Knoll, 2003).

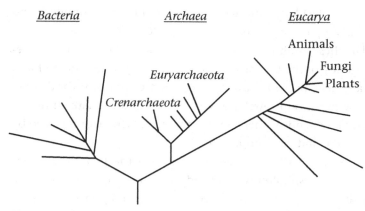

Figure 14.1. *The universal phylogenetic tree determined by sequencing of SSU rRNA.* Reprinted with permission from C. R. Woese, Interpreting the universal phylogenetic tree, *Proc. Natl. Acad. Sci. USA*, **97**, 8392–8396. Copyright 2000 National Academy of Science USA.

evidence was taken to indicate that advanced microbial life was already present on Earth 3.5 billion years ago, perhaps even as early as 3.8 billion years ago (Schopf, 1999). The significance of this evidence has since been questioned (Brasier et al., 2002; van Zullen et al., 2002; Garcia-Ruiz et al., 2003), but enough independent signs remain that life does indeed go back to at least 3.5 billion years ago (Schopf, 1999; Knoll, 2003; Furnes et al., 2004).

The Great Prokaryote Split

Woese's discoveries have had a considerable impact on our appreciation of the prokaryotic world. Whereas Bacteria were found to comprise most of the organisms classically studied by bacteriologists and microbiologists, Archaea turned out to include many previously unknown denizens of soils and oceans, adapted to a remarkable variety of environments. Among them are the widely distributed methanogens, which inhabit many anaerobic milieus, where they survive by converting hydrogen and carbon dioxide to methane and water; and the so-called extremophiles, adapted to particularly harsh environments, including extreme temperature, up to

110°C (thermophiles),[4] extreme acidity, down to pH 1 (acidophiles), and extreme salinity, up to saturated brine (halophiles).

Sparked by these findings, a considerable amount of work has been devoted to the study of the two prokaryotic groups, especially the Archaea, of which many new species have been discovered. Confirming the distinctiveness of the two groups, Archaea have been found to share a number of genetic and biochemical characteristics that are lacking or replaced by others in Bacteria, and vice versa.

One of the most striking differences between the two groups concerns membrane phospholipids. As already seen in Chapter 10, bacterial phospholipids contain fatty acids linked to L-glycerol 3-phosphate by ester bonds, whereas the archaeal molecules contain isoprenoid alcohols linked to D-glycerol 3-phosphate by ether bonds.

Another characteristic difference between Archaea and Bacteria relates to the cell wall, a rigid, protective structure that surrounds many prokaryotes. In Bacteria, the cell wall is made of a complex substance, called murein, in which sugars and both L- and D-amino acids are cross-bridged together into a continuous network that completely surrounds the cell. The cell wall of Archaea is not made of murein; it sometimes consists of a similar substance, pseudomurein, containing only L-amino acids.

Faced with the existence of the two prokaryotic domains, one wonders what might have caused an initial divergence decisive enough to create two sharply distinct branches capable each of fanning out into a wide variety of forms. Considering the clustering of extremophiles among Archaea, it is tempting to assume that adaptation to a harsh environment, most likely excessive heat, was the selective criterion, separating the more resistant archaean ancestor from the more fragile founder of the bacterial line, relegated to milder surroundings. This hypothesis is supported by the fact that the most ancient archaean organisms are all thermophilic. True, the more ancient bacteria are also thermophilic, but their heat tolerance is lower than that of the most resistant archaean thermophiles. In addition, there is evidence that thermophilic bacteria may have received a

---

[4] A recent work has shown that this limit could exceed 120°C and even reach 130°C (Kashefi and Lovley, 2003).

number of genes from archaean thermophiles by lateral transfer (Aravind et al., 1998; Nelson et al., 1999; Forterre et al., 2000).

Heat tolerance depends on a number of factors, including the ether membrane phospholipids already mentioned, a particularly high G+C content of rRNAs and tRNAs (the pairing bond between G and C is stronger than that between A and U), and a greater resistance of the proteins to heat denaturation, a property that is now exploited on a large scale for the industrial production of heat-resistant enzymes. The powerful technology known as PCR (polymerase chain reaction), which is universally used for the selective isolation and amplification of DNA, involves a heating step that has been greatly facilitated by the availability of a heat-resistant enzyme extracted from thermophilic organisms.

From a theoretical point of view, it would seem that adaptation to heat should be more easily lost than gained: loss of a single, critical property suffices to abolish heat tolerance, whereas improved heat tolerance may depend on the concerted acquisition of a number of different, unrelated properties. This kind of reasoning suggests that Archaea segregated originally by virtue of their resistance to excessive heat and that they inherited this property directly from a thermophilic LUCA, leaving Bacteria to emerge later, when a milder environment became accessible.

We have seen that there are good reasons to believe that the LUCA may have been the product of a selective bottleneck. Furthermore, the thermophilic character of all the most ancient organisms makes it tempting to assume that excessive heat was responsible for this bottleneck. Also supporting the hypothesis of a thermophilic LUCA is the implicit notion that this organism possessed ether membrane phospholipids, a possibility consistent with the strongly documented proposal by Ourisson and Nakatani (1994) that terpenoid derivatives, rather than fatty acids, provided the hydrophobic chains of the first membrane phospholipids. Whether heat resistance itself may have been acquired during the long pre-LUCA evolutionary process or may go back all the way to a "hot cradle" is a question beyond even educated guesswork.

The hypothesis that Archaea arose directly from a thermophilic LUCA is not unanimously accepted. The French investigator Patrick Forterre, in particular, has championed the theory that thermophilic archaeans, as well as thermophilic bacteria, have evolved from a mesophilic (adapted to a mild temperature) LUCA by a process of "thermoreduction" (Forterre, 1995, 1999; Xu and Glansdorff, 2002). In support of this theory, a phylogenetic study based on the G+C content of rRNA has also led to the conclusion that the LUCA was mesophilic (Galtier et al., 1999).

The Protoeukaryote

The most contentious aspect of phylogenetic studies concerns the ancestry of eukaryotes. Early results (see Figure 14.1) suggested that the first fork led to the separation of Archaea and Bacteria and that eukaryotes diverged later from the archaeal line. As investigations were extended to a larger number of genes, however, some eukaryotic genes appeared to be related to archaeal genes, in agreement with the early findings, but others were found to be closer to bacterial genes. Thus, the reconstructed genome of the *protoeukaryote*, the founder of the eukaryotic domain (also called "urkaryote" by Woese), seems to be a chimera of archaean and bacterial genes.

The mixture is not haphazard. The eukaryotic genes of archaeal character are mostly concerned with replication, transcription, and translation. On the other hand, the eukaryotic genes of bacterial character deal mostly with metabolism and other "housekeeping" tasks. The former are informational, the latter operational (Lake et al., 1999). A particularly striking similarity between eukaryotes and Bacteria concerns membrane phospholipids. Eukaryotes, like Bacteria, use ester phospholipids for the construction of their membranes, instead of the ether phospholipids one would expect if they originated from Archaea.

A great variety of explanations have been proposed to account for the genetic chimerism of the protoeukaryote, with hardly any consensus emerging so far. The field is in constant turmoil, and I cannot possibly do justice to it within the limits of the present book. I can do no more than mention some of the issues, with a few selected references that should help the interested reader delve deeper into the subject.

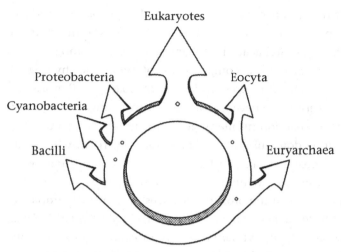

Figure 14.2. *The ring of life*. Representation of the origin of eukaryotes by genome fusion between a bacterial and an archaean ancestor. Bacteria are on the left, Archaea on the right. The root is believed to be situated below, on the bacterial side. Reprinted with permission from M. C. Rivera and J. A. Lake, The ring of life provides evidence for a genome fusion origin of eukaryotes. *Nature*, **431**, 152–155. Copyright 2004 Nature magazine.

A first group of explanations attributes the observed chimerism to some kind of merger, by cell fusion, symbiosis, or endosymbiosis,[5] leading to a hybrid cell endowed with an archaean nucleus and a bacterial cytoplasm (Zillig, 1991; Gupta et al., 1994; Margulis, 1996; Horiike et al., 2001). Strong phylogenetic support for the genome-fusion theory has been provided in a recent paper by Rivera and Lake (2004), who have pointed out that if two ancestors, of archaean and bacterial origin, respectively, joined to make the first eukaryote, the "tree of life" is no longer a tree but a ring (see Figure 14.2). The authors do not, however, offer any explanation for the mechanism whereby archaean (informational) and bacterial (operational) genes were selectively retained in the alleged fused genome.

[5] As will be explained in the next chapter, the term "endosymbiosis" refers to a process whereby a cell is taken up by another cell, progressively evolves *within* the latter, and becomes subservient to it, up to, sometimes, turning into an organelle. *Symbiosis*, on the other hand, designates a relationship between two cells or organisms that both derive an advantage from living together, while preserving their individuality. We shall see that the transition from symbiosis to endosymbiosis is a hot topic of contemporary discussion.

A variant of the fusion proposal, known as the *syntrophic hypothesis* (Moreira and Lopez-Garcia, 1998), envisages the nucleus of the chimera as originating from an archaean, hydrogen-utilizing methanogen, and its cytoplasm as formed by a conglomerate of symbiotic, hydrogen-producing bacteria, related to present-day δ-proteobacteria, that assembled around the central archaean, coalesced into a single entity, lost their nuclei (while abandoning many genes to the archaean nucleus), and developed an endomembrane system. Another variant, which will be considered in greater detail in Chapter 17, also rests on hydrogen-dependent symbiosis but identifies the bacterial participant in this hybridization process with the prokaryotic ancestor of mitochondria, allegedly engulfed by an archaean host cell (Martin and Müller, 1998; Vellai and Vida, 1999) Yet another variant assumes a distinct protoeukaryote that acquired its archaean and bacterial genes by two separate endosymbiotic events (Hartman and Fedorov, 2002).

Another kind of interpretation invokes lateral gene transfer as the main cause of chimerism. As briefly explained before, lateral, or horizontal, gene transfer takes place between cells that may be entirely unrelated. Now well substantiated, this phenomenon can happen by a variety of mechanisms and has been found to occur on a large scale in the prokaryotic world (Aravind et al., 1998; Nelson et al., 1999; Forterre et al., 2000; Ochman et al., 2000) and even among unicellular eukaryotes (Andersson et al., 2003). Cogently defended by Woese, the founder of the field (Woese, 1998, 2002), and by the Canadian investigator Ford Doolittle (1999, 2000), this theory has been adopted by a number of other workers, to the point, as will be seen below, of blurring the very concept of LUCA. As illustrated in Figure 14.3, frequent enough lateral gene transfers would transform the "tree of life" into a reticulated form, or net.

Yet another theory, advocated by, among others, the Belgian researcher Nicolas Glansdorff (2000), puts the main onus on genetic paralogy. This phenomenon is the consequence of gene duplication leading to the formation of two (paralogous) copies of the same gene, which can then have distinct evolutionary histories. There is considerable

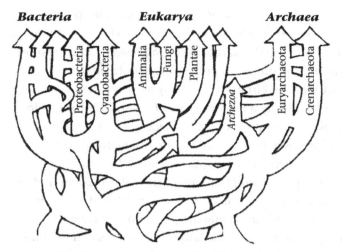

Bacteria   **Eukarya**   **Archaea**

Figure 14.3. *The reticulated tree of life.* Multiple lateral gene transfers convert the tree into a network. Reprinted with permission from W. F. Doolittle, Phylogenetic classification and the universal tree, *Science*, **284**, 2124–2128. Copyright 1999 AAAS.

evidence that such duplications occurred even before the emergence of the LUCA.

It does not behoove me, as a mere onlooker attempting, with a focus on singularities, to unravel the tangled skein of facts and surmises offered by the rich literature on the topic, to take a stand on such a controversial matter. I shall restrict myself to a few remarks of a general nature.

First, it seems to me that Woese's rRNA-based phylogeny, which he himself does not hesitate to qualify as "canonical," is likely to be correct. This is not just a case of beginner's luck, but the outcome of a perceptive choice. Ribosomal RNAs go back to the very beginnings of life, they are highly conserved, and, as parts of a complex interlocked machinery of fundamental importance, they are likely to have been transferred exclusively by direct, vertical inheritance. Subsequent circularization or reticulation of the tree should not obscure this basic relationship.

A second point is that it does not seem possible, in the present state of knowledge, to put dates, even approximative ones, on the key bifurcations of the canonical tree. Too many hazards weigh on the interpretation of evolutionary distances, especially very ancient ones. This uncertainty is particularly great as concerns the start of the lineage leading to eukaryotes.

Coming to the actual chimerism problem, it should not be forgotten that eukaryotes are separated from the LUCA by a history that is not only very long but also marked by a large number of remarkable innovations that did not take place in the prokaryotic lineages and that have involved at least one kind of endosymbiotic event, a source of many foreign genes. This question will be examined in Chapters 15 and 17.

Finally, it is true that, as emphasized by Glansdorff (2000) and by Wächtershäuser (2003), among others, special attention deserves to be given to the sharp difference in membrane phospholipid chemistry that separates Bacteria and Eucarya, on one hand, and Archaea, on the other. The difference goes back to two chirally distinct pathways, presumably initiated at the level of dihydroxyacetone phosphate reduction,[6] and it could well represent a typical case of enforced homochirality, as membranes containing the two kinds of lipids are probably unstable (a point, incidentally, that could be studied experimentally). This raises the problem how one kind of phospholipid could have replaced the other in membranes, as must have occurred at least once, perhaps twice. Of possible significance in this connection is the fact that the ester phospholipids share a joint origin with neutral lipids, which are not constituents of membranes, serving only for energy storage. Both originate from L-glycerol 3-phosphate. In a cell line making neutral lipids, a small genetic change would have sufficed to bring about the formation of ester phospholipids. Acquisition of this ability by horizontal gene transfer or even by convergent evolution, is thus readily conceivable and could have taken place more than once, letting natural selection tip the balance in favor of one or the other homochiral membrane structure, depending on environmental conditions (Wächtershäuser, 2003).

---

[6] A recent phylogenetic study (Boucher et al., 2004) has indeed shown that the enzyme catalyzing the formation of D-glycerol 3-phosphate from dihydroxyacetone phosphate [see Chapter 10, equation (2)] is one of the few enzymes of ether lipid synthesis that is exclusively archaeal. A number of other enzymes vary among archaeans or are related to their bacterial homologues in a manner indicating multiple lateral gene transfers.

Reappraising the LUCA

The considerations mentioned above have greatly affected the representation of the LUCA, which defenders of lateral gene transfer now tend to depict as some sort of community of primitive cells that exchanged genes so freely among each other as to completely lose their individuality. In the words of Woese (1998, p. 6858), "the ancestor cannot have been a single organismal lineage. It was communal, a loosely knit, diverse conglomeration of primitive cells that evolved as a unit, and it eventually developed to a stage where it broke into several distinct communities." Others have advocated similar views of the LUCA (Kandler, 1994a, 1994b; Doolittle, 1999; Martin, 1999; Koonin, 2003).

The issue hinges largely on the significance one gives to the word "last" in the phrase "last common ancestor." As understood in the present book, this term refers strictly to the root of the phylogenetic tree, that is, the living form that, for the first time in the development of life, branched into two distinct lines that both extend to the present day. For the defenders of a communal ancestor, the term encompasses the whole of the long history, recalled in the preceding chapters, that ends with the appearance of the last ancestor. Thus, Woese (1998) refers to primitive cell forms, or "progenotes," going back almost to the RNA world and equipped with a simple collection of small genes (apparently of DNA nature, however) subject to rudimentary replication, transcription, and translation, which he imagines as slowly progressing, much in the manner outlined in Chapter 8, toward genes and proteins of increasing length by a process bootstrapped by the acquisition of replicating systems of greater accuracy. Progenotes are depicted as freely exchanging genes all along this process, in some kind of "common ancestral chaos." In a similar vein, Kandler (1994b) speaks of a "multiphenotypical population of pre-cells," exhibiting "frequent mutual exchanges of templates."

Regardless of semantics, it seems to me that the "communal" picture of the LUCA lacks the mainspring of evolutionary progression, *selection*. Collectivism is the antithesis of competition, and competition is the essence of Darwinian selection. The same objection has been made by Forterre, who is quoted as having said: "to think of LUCA in terms of community is to remove the idea of Darwinism from early evolution" (Whitfield, 2004). It is a striking feature of the papers proposing the new view of the LUCA that they seem to exclude a role of selection in early evolution. Thus, we are told that "from the start the course of cellular evolution is a march toward greater complexity, integration, precision, specificity of cellular design, etc.," that this is "a task that can be accomplished only by collective evolution in which many diverse cell designs evolve simultaneously and share their novelties with one another" (Woese, 2002, p. 8745). But the driving force behind the march, the mechanisms whereby useful novelties were separated from those (no doubt the majority) that were either useless or deleterious, are not mentioned.[7]

In contrast, the nature of these mechanisms has been the almost tiresomely repeated *leitmotiv* of this book. Selection, we have seen, must have emerged at the same time as replicable RNA. After a brief spell when it operated directly at the molecular level, this process must soon have become indirect and based on the *usefulness* of the genes or of their products. The whole developmental history of protein synthesis and progressive lengthening, of the genetic code, of the enzymes that came to catalyze the first reactions of metabolism, and of all the other improvements that led to the LUCA was, according to the reconstruction proposed in this book, dominated by selection processes that mandated the participation of competing protocells endowed with distinctly more stringent genetic individuality than characterizes the hypothesized progenotes or pre-cells (de Duve, 2005). Most likely, these protocells did engage in a certain amount of gene swapping by lateral transfer, but they cannot

[7] The omission seems to be deliberate, as indicated by the following statement by Woese: "In the modern world, evolutionary innovation tends to become established through selection acting on organisms, whereas in a world dominated by lateral gene transfer, an innovation takes over by direct 'invasion'" (Woese, 1998, p. 6858). The two regimes are pictured as separated by what Woese (2002) has called the "Darwinian Threshold," a "critical point" in the evolution of the progenote that marks a "phase change" from the communal state to that of a more integrated cellular organization. Crossing this threshold is presented as a very late event, posterior to the separation of Bacteria, which crossed it first, from Archaea and Eukarya, which did so later (Woese, 2002).

have done so to the extent alleged unless early evolution was driven not by natural selection but by some other mechanism that remains to be identified.[8]

[8] A possible way out of the dilemma is provided by Woese in a recent overview with strong historical and philosophical overtones, in which he writes: "Novelty, of course, doesn't exist in a vacuum; it has to have selective value in some environment. For this reason, I see no way out of the conclusion that cellular evolution began in a highly multiplex fashion, from many initial independent ancestral starting points, not just a single one. Such a strategy automatically optimizes both the amount and diversity of novelty generated because it generates a great variety of selective contexts" (Woese, 2004, p. 183). Applying this rather cryptic pronouncement to, for example, the mechanism of enzyme selection discussed in Chapter 3 suggests the possibility that different enzymes could have arisen simultaneously in different protocell lines, rather than serially in a single line, subsequently spreading to other lines by lateral transfer of the genes concerned. This possibility does, however, imply that lateral gene transfer among early protocells was by far not as pervasive as is intimated by its advocates; it must, like its modern counterpart, have been sufficiently infrequent to allow different lines to follow distinct evolutionary histories, individually traced by natural selection.

15 Eukaryotes

The advent of eukaryotic cells represents one of the great, epoch-making occurrences in the history of life on our planet. Through this event, the biosphere, which had long remained composed exclusively of simple prokaryotes, became enriched with a wealth of new microorganisms considerably more complex than the first ones and, eventually, with all plants, fungi, and animals; without it, we wouldn't be around.

When this fateful event took place is not precisely known. As we saw in the preceding chapter, there is evidence, provided largely by ribosomal RNA, that the eukaryotic line started diverging from the prokaryotic ones early after LUCA days – that is, 3.5 billion years ago – or perhaps even before. But the time of appearance of the first typical eukaryotic cells remains highly uncertain. The oldest clearly identified eukaryotic fossils are dated 1.5 billion years ago (Knoll, 2003). Earlier traces have been described, but their eukaryotic nature has been disputed. An origin going back to as early as 2.7 billion years ago is suggested by the presence, in terrains of that age, of steranes, the parent group of cholesterol, steroid hormones, and other derivatives, which are universal constituents of eukaryotes, though not entirely exclusive ones (Brocks et al., 1999). This thus leaves, between the LUCA and the earliest known eukaryotes, a gap of at least 800 million years on which the fossil record is entirely silent. All we have to fill this gap is the information extant organisms can offer on their history. As we shall see, the clues are far from negligible but still very incomplete and open to conflicting and hotly debated interpretations.

What we do know with a considerable degree of confidence, however, is that this long, mysterious road ended in a singularity; it produced a new type of cell endowed with a whole series of characteristic properties, ancestral to all present-day eukaryotes, be they unicellular or multicellular. Before considering the origin of these new cells, let us first look at the principal features that distinguish them from prokaryotic cells.

The Hallmarks of Eukaryotic Cells

Eukaryotic cells are, like prokaryotic cells, membrane-bounded entities containing an enzyme-rich cytosol, in which, among other components, the protein-synthesizing ribosomes are situated. But eukaryotic cells differ from their prokaryotic relatives by a number of characteristic traits, shown schematically in Figure 15.1: (1) a much larger size (20 μm in diameter, on average, as against 1 μm, which amounts to an almost ten-thousand-fold increase in volume); (2) a surrounding membrane much more richly and selectively specialized to carry out exchanges of matter (transporters) and information (receptors) with the outside; (3) an internal cytomembrane system functioning in the uptake (endocytosis) and digestion (lysosomes) of extracellular materials (import) and in the manufacture [endoplasmic reticulum (ER), Golgi apparatus] and discharge (exocytosis) of secretion products (export); (4) a cytoskeleton (consisting mainly of actin fibers and microtubules), connected with ATP-powered motor elements (myosin, dynein, kinesin); (5) highly structured chromosomes segregated, together with the main systems of DNA replication, DNA transcription, and RNA processing, inside an enclosure (nucleus) separated from the rest of the cell (cytoplasm) by a membranous envelope (nuclear envelope) reinforced by cytoskeletal components (lamina) and pierced by pores that allow selective exchanges with the cytoplasm; and (6) membrane-bounded organelles involved in oxygen consumption (peroxisomes, mitochondria) and, only in unicellular algae and green plants, in oxygen-yielding photosynthesis (chloroplasts). In addition, (7) eukaryotic cells divide by a complex mechanism (mitosis)

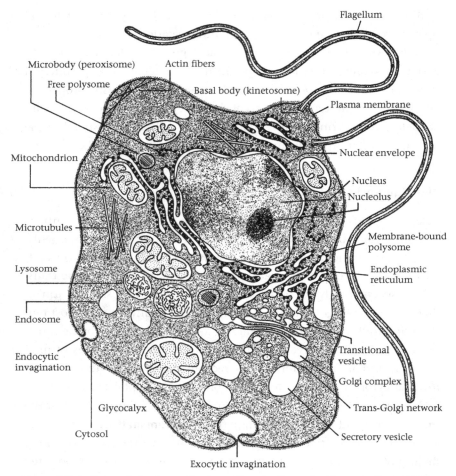

Figure 15.1. *The main features of eukaryotic cells.* Excerpted from de Duve (1991, p. 56), the picture is that of a hypothetical protist displaying all the components of modern eukaryotic cells, with the exception of chloroplasts, which are present only in unicellular algae and in plant cells.

dependent on the construction of a specialized scaffolding (spindle) made of microtubules.

As can be found in any modern cell biology textbook, all this is common knowledge. The structures and functions of the various cell parts are now understood in great detail and need not be considered further in this chapter, which will be devoted only to origins and singularities.

The Origin of Eukaryotic Cells

It is commonly accepted that the main features of eukaryotic cells were acquired after the appearance of the LUCA, by an entity that had, essentially, a prokaryotic type of organization. This opinion is no longer unanimously shared today. As we have seen in the preceding chapter, some investigators imagine a LUCA closer to eukaryotes, of which prokaryotes would be descendants rather than ancestors. Some even push back the origin of eukaryotes to the RNA world.

These notions are based on phylogenetic considerations and are rarely associated with attempts to explain how the characteristic properties of eukaryotes may have been acquired. The magnitude of this task can be assessed by a glance at Figure 15.1. What has to be accounted for is the formation of the whole, complex, depicted structure, starting from an ancestral form that, even if not comparable to present-day prokaryotes, must in any case have shared with them a rudimentary cellular organization. For this reason, I shall, in the discussion that follows, adopt as starting hypothesis that eukaryotic cells have arisen, as assumed by the classical hypothesis, from ancestral entities of prokaryotic type. There will always be time later on to see how the proposed models can be adapted to different ancestral forms.

The mechanisms involved in the famous prokaryote–eukaryote transition have long been subject to abundant speculations, especially after the decisive advances of cell biology in the 1950s. But the question truly got off the ground only after 1967, when the American biologist Lynn Margulis, under the name of Lynn Sagan (she was at that time married to the celebrated physicist and media personality Carl Sagan), revived the old but abandoned theory of the prokaryotic origin of mitochondria and chloroplasts (Sagan, 1967). First received with skepticism, this theory has now reached the status of established fact. The evidence is overwhelming that mitochondria and chloroplasts are descendants from erstwhile free-living prokaryotes that were adopted as *endosymbionts* – literally, living together inside – by certain host cells. A similar origin of peroxisomes has also been envisioned, but this possibility lacks confirmation or, for that matter, refutation.

Much has been learned of the nature of the prokaryotic ancestors of eukaryotic organelles, their endosymbiotic adoption, and their integration within the economy of their host cells, all topics to be addressed in Chapter 17. But the nature of the host cells themselves and the mechanisms whereby they came to harbor the endosymbionts remain matters of considerable uncertainty and debate. In essence, two main theories have been proposed. They can be described under the labels of *primitive phagocyte* and *fateful encounter*.

The Primitive-Phagocyte Hypothesis

Even before the endosymbiont theory was proposed, workers had reflected on the origin of some of the eukaryotic features brought to light by the new techniques of electron microscopy and tissue fractionation. In my own research, I had become particularly intrigued by the origin of lysosomes, which had recently been recognized as digestive organelles seemingly present in all eukaryotic cells and involved in the breakdown of extracellular materials taken up by endocytosis and in that of intracellular materials segregated by autophagy (de Duve, 1963). In a review written in collaboration with my Belgian coworker Robert Wattiaux (de Duve and Wattiaux, 1966), we made the proposal illustrated in Figure 15.2. As a starting assumption, we posited the existence of a wall-less, heterotrophic prokaryote that depended, like similar present-day organisms, on the extracellular digestion of food by secreted exoenzymes. It all started in this organism, we suggested, with a genetic change that caused its peripheral membrane to expand excessively. As a necessary consequence of this phenomenon, the membrane would have folded into increasingly deep invaginations, sometimes leading, through simple fusion, to the formation of intracellular vesicles containing trapped extracellular materials, prefiguring endocytosis. Food particles dragged in by this phenomenon would be digested inside the vesicles by the exoenzymes, which would continue to be discharged co-translationally across the membrane, but ending up inside the vesicles instead of outside the cell. Originally combining the properties of endosomes, rough endoplasmic reticulum cisternae, and lysosomes, the postulated vesicles

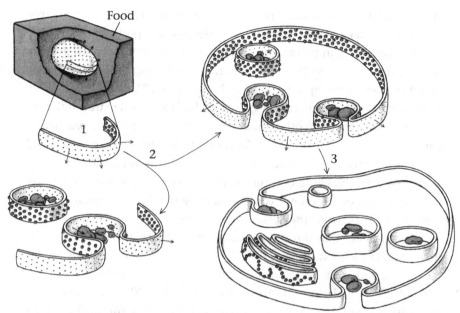

Figure 15.2. *Hypothetical mechanism explaining the evolutionary origin of a primitive phagocyte from a prokaryotic ancestor:* **(1)** like many present-day heterotrophic prokaryotes, the putative ancestor relies for nutrition on the extracellular digestion of food by exoenzymes that are made by plasma-membrane-bound ribosomes and discharged outside; **(2)** invaginations of the plasma membrane in a wall-less organism create primitive intracellular digestive pockets in which interiorized food particles (primitive endocytosis) are broken down by trapped exoenzymes (primitive lysosome); **(3)** differentiation of the intracellular membrane system relegates bound ribosomes to parts of the endoplasmic reticulum and leads to the formation of distinct membranous domains, and the cell increases considerably in size. Illustration from de Duve (1984, p. 98).

would subsequently have differentiated into the various parts of the eukaryotic cytomembrane system.[1]

There are attractive aspects to this model: the postulated process could have depended on a simple genetic change, leading to enhanced phospholipid synthesis, for example; and, especially, it includes a powerful, plausible mechanism for its own selection. As we wrote, "the advantages of this development are obvious. It relieved heterotrophy from the ecological requirement for a confined and relatively stagnant environment;

---

[1] For a more detailed description of the model, see de Duve (1995, 1996, 2002).

at the same time, it made larger membrane areas available for nutritive exchanges to take place efficiently with deep-seated portions of the cell, possibly providing in this manner an opportunity for further cell growth and organization" (de Duve and Wattiaux, 1966, pp. 471–472). I conveyed the same message later in more lyrical terms: "Until then, in order to benefit from their exoenzymes, the cells had to rely on extracellular digestion. Unless they had other means of subsistence, they were practically condemned to reside inside their food supply, like maggots in a chunk of cheese. Henceforth, they would be free to roam the world and to pursue their prey actively, living on phagocytized bacteria or on other engulfed materials. This development could well have heralded the beginning of cellular emancipation" (de Duve, 1984, p. 99).

A similar idea was put forward independently by the microbiologist Roger Stanier, who wrote in 1970: "The capacity for endocytosis would have conferred on its early possessors a new biological means for obtaining nutrients: predation on other cells" (Stanier, 1970, p. 27). The most committed advocate of the theory was – and still is – the British biologist Thomas Cavalier-Smith (2002), who made the same suggestion, also independently, adding the important point that the development of the membrane system must have gone together with the coevolutionary development of the cytoskeletal and motile elements needed to direct the membrane movements. In his words, written in 1975 (when he assumed that the ancestral cell was a cyanobacterium), "a phagocytic pre-alga could photosynthesize by day or during the Arctic summer, and phagocytose by night or during the Arctic winter (or in any dark environment)" (Cavalier-Smith, 1975, p. 463). The images differ, but the main idea is the same.

It must be emphasized that our model *antedates* the endosymbiont hypothesis; its objective was to account for the development of some of the main features of eukaryotic organization, in particular lysosomes, which indeed it does. Remove the oxygen-consuming organelles from Figure 15.1, and you are left with an entity that could conceivably have arisen by the proposed mechanism. The name "primitive phagocyte" was given to this hypothetical ancestral form because ability to feed by phagocytosis and to digest the captured food intracellularly was seen as the cardinal feature of the organism. Several arguments support the model.

First, there are the many indications of kinship between eukaryotic cytomembranes and prokaryotic cell membranes. The two harbor many phylogenetically related systems, including protein-translocating mechanisms that go so far as to recognize each other's signal sequences. There is thus ample proof that, as surmised, the eukaryotic cytomembrane system arose by expansion and internalization of a prokaryotic cell membrane. A side aspect of the model is that it could even account for the birth of the eukaryotic nucleus. It is known that, in prokaryotes, the chromosome is anchored to the cell membrane. If the membrane patch involved in this anchoring had become internalized in the manner suggested, it could have ended inside the cell in the form of a membranous sac containing the chromosome, the most primitive form of a nucleus (de Duve, 1991). A possible vestige of this process is the fact that, in present-day cells, the nuclear envelope is assembled from endoplasmic reticulum vesicles.

Given the prokaryotic origin of eukaryotic membranes, a significant advantage of the proposed model is that it accounts for the implied transition by a smooth succession of plausible, viable intermediates, with the passage from extracellular to intracellular digestion as the guiding thread. The only major innovations required – this is true of any hypothesis one may propose – concern some of the cytoskeletal proteins, which must have arisen from ancestral prokaryotic molecules that still await accurate identification (Nogales et al., 1998; van den Ent et al., 2001).

The greatest merit of the primitive-phagocyte hypothesis is that it allows for a continual, powerful, selective support of the assumed transformation. There can be little doubt that the ability to capture food and to digest it intracellularly must represent an immense asset for any unicellular, heterotrophic organism. Furthermore, the selective pressure could have been maintained uninterruptedly during what must obviously have been a long, gradual process, because every step would have entailed an evolutionary benefit.

No other mechanism has, so far, been proposed to explain these crucial developments. So the primitive-phagocyte hypothesis deserves consideration whatever mechanism one envisions for the adoption of endosymbionts. As it turns out, however, the hypothesis seems almost made to order to account for this phenomenon. Phagocytosis is the universal mechanism whereby eukaryotic cells capture microorganisms. In most cases, the captives are killed and digested within lysosomes; they occasionally escape this fate and develop either inside vesicles of the cytomembrane system or in the cytosol, usually leading to sickness and death of the affected cells. Exceptionally, host and prey establish a relationship of mutual tolerance that tends to become symbiotic to the extent that each partner of the relationship provides an asset to the survival of the other. Many such examples are known. Given enough time, the progressive enslavement of the endosymbiont by the host cell, notably by way of massive gene transfers, is readily visualized (see Chapter 17).

The case for the primitive-phagocyte hypothesis received what appeared as clinching support from the first molecular phylogenies, which pointed to several groups of microorganisms devoid of mitochondria, including diplomonads, microsporidia, and trichomonads, as occupying the lowest branches of the eukaryotic tree. Here, it was hailed, were descendants of the primitive phagocyte that branched from the parent line before the endosymbiotic adoption of mitochondria.

This explanation was later severely shaken by various observations[2] that suggested strongly that a number of amitochondriate (without mitochondria) protists may once have contained mitochondria and lost them subsequently in the course of evolution. Several genes believed to be of mitochondrial origin were identified in the organisms' nuclear genomes, into which they had presumably been incorporated by transfer from the endosymbiont before it was lost (see Chapter 17). Small membrane-bounded structures, taken to be mitochondrial remnants and called *mitosomes* for this reason, were even recognized in the cytoplasm of some of the organisms. One of these organisms, namely a microsporidium, turned out, in addition, to be badly misplaced on the early tree, actually belonging to the group of fungi. Finally, there is the evidence, to be examined

---

[2] A detailed account of this important topic is beyond the scope of this book. Interested readers will find additional information and most of the relevant bibliography in Martin and Müller (1998), Martin et al. (2003), and Tovar et al. (2003).

in greater detail in Chapter 17, indicating that some organelles, known as *hydrogenosomes*, may be related to mitochondria. Because of these various findings, it is widely believed, nowadays, that there is no eukaryotic organism, among those that have been adequately explored, whose ancestry does not include a forebear that once contained mitochondria or a related organelle, the inference being that, if the primitive phagocyte ever existed, it has left no extant descendant of a lineage that never contained endosymbionts. In other words, the adoption of mitochondria signals the eukaryotic singularity. In the eyes of many observers, these facts invalidate the primitive-phagocyte hypothesis, a "view founded more in tradition than in evidence" (Martin et al., 2003).

Thus, one might say, quoting T. H. Huxley, was a beautiful hypothesis killed by an ugly fact. "Grossly overkilled" is really what should be said. It is obvious that the new findings offer no reason for rejecting a hypothesis that has in no way lost verisimilitude but only lost what was mistakenly taken to be a confirmation. In the words of Carl Sagan, absence of evidence is not evidence of absence. We should remember that the world of protists is still poorly known. Perhaps some surviving descendant of the primitive phagocyte is still lurking somewhere, waiting to be discovered.[3]

It is also worth recalling that the primitive-phagocyte hypothesis was not proposed to account for endosymbiont adoption but to explain how some of the main eukaryotic features, especially the cytomembrane system and the cytoskeleton, could have developed in the course of evolution. The fact is that no other plausible hypothesis has yet been put forward to account for these momentous changes; they are either ignored or glibly covered in a single sentence that offers no explanation, whether mechanistic or selective.

Mention must be made here of a recent study by Hartman and Fedorov (2002), which has enriched the field with new data of great interest. Comparing a set of 2136 protein sequences found in databases, the authors have singled out a total of 347 proteins "that are found in eukaryotic cells

---

[3] Interestingly, membranous structures and what could be the inception of a nucleus have recently been described in microbes called planctomycetes (Lindsay et al., 2001). I am indebted to Nicolas Glansdorff for having drawn my attention to this interesting observation (which, incidentally, is mostly cited in support of a eukaryotic LUCA).

but have no significant homology to proteins in Archaea and Bacteria." A number of these proteins are connected with the cytomembrane and cytoskeletal systems. The authors interpret their findings as evidence of the early existence of a phagocytic cell ancestral to all eukaryotes. Called the *chronocyte*, this ancestral cell is taken to have separated from prokaryotic lines in the RNA world and to have continued functioning with an RNA genome until engulfment of an archaean prokaryote endowed it with an endosymbiotic, DNA-based nucleus.[4]

This view, which was proposed many years ago by Hartman (1984) and has since been adopted also by the Sogin group (Sogin et al., 1996), is even more radical than Woese's model of the LUCA referred to in the preceding chapter (Woese, 1998). Should it be verified, the model proposed for the chronocyte not only would demand a complete reappraisal of our views on the LUCA and the first fork but also would cause us to revise many of our ideas concerning the evolutionary potential of RNA genes. The arguments offered until now in support of the model are, however, far from being sufficiently compelling to justify such an upheaval.

Note that the actual data are not particularly surprising in themselves, for it is evident that the development of eukaryotes must have involved many innovations that did not take place in the prokaryotic lines. What may seem surprising and represents the authors' main argument is the lack of obvious kinship between the sequences under consideration and prokaryotic sequences. Given the enormous durations involved – at least 800 million years, as we have seen – and the fact that innovations may diverge more rapidly from a parent line than mere functional continuations, the possibility that the proteins share ancestry with prokaryotic proteins but that traces of this ancestry have largely been erased cannot be ruled out. Remember that vestigial traces of a prokaryotic origin have already been found for certain typical eukaryotic proteins (Nogales et al., 1998; van den Ent et al., 2001); others may well be uncovered in the future.

Even if such allowances are made, the fact remains that the proteins concerned must be very ancient, especially in that many distinct species

---

[4] The model of an RNA-based protoeukaryote proposed by Poole et al. (1999) deserves to be recalled in this connection (see footnote 2, Chapter 8).

are involved. As will be pointed out below, such antiquity by indicating that the main features of eukaryotic cells were acquired early, argues in favor of the primitive-phagocyte hypothesis. On the other hand, the claim that this cell may have functioned on an RNA genome rests so far on flimsy evidence and will need much stronger corroboration before it is accepted. Quoting Carl Sagan again, extraordinary claims require extraordinary evidence.

The Fateful-Encounter Hypothesis

This hypothesis was proposed by Lynn Margulis in her historic article on the endosymbiotic origin of organelles (Sagan, 1967). In that paper, and in later works, she depicted the postulated encounter as an attack on some unidentified prokaryote, which became the host cell, by a "fierce predator," also of prokaryotic nature and tentatively identified as a *Bdellovibrio* relative, destined to become the endosymbiont (Margulis, 1981; Margulis and Sagan, 1986). She dismissed the alternative hypothesis of an endocytic uptake of the endosymbiont on the ground that "pinocytosis and phagocytosis have never been seen in prokaryotes" (Margulis, 1981), and she suggested, instead, that it was the acquisition of mitochondria that allowed the development of phagotrophy. She did, however, credit phagocytosis for the subsequent uptake of the ancestors of chloroplasts. As to the other eukaryotic features, her main focus has been on what she calls "undulipodia," the motor appendages, characterized by the same arrangement of microtubules, that comprise eukaryotic cilia and flagella. These appendages, she has claimed, were acquired by another fateful encounter, involving a relative of present-day spirochetes, motile prokaryotes that include the agents of syphilis and of Lyme disease. This hypothesis, which confuses two entirely unrelated types of flagella, has few adherents today.

It is a tribute to the authority Margulis has gained as the champion of the endosymbiont theory – and also a tribute to her charismatic persuasiveness – that the fateful-encounter model has found wide acceptance in spite of its being largely gratuitous. The theory has even gained greater credit in recent years with the findings, reported in the preceding chapter, suggesting that the eukaryotic genome is a chimeric association of an archaeal and a bacterial genome. Defenders of this fusion theory all take the phenomenon as initiated by an encounter between two prokaryotes leading to an endosymbiotic relationship. A particularly persuasive case in favor of the fateful-encounter theory combines genome fusion with the origin of mitochondria (Martin and Müller, 1998). Remarkably, none of the defenders of this theory elaborates on the alleged pathway leading from a symbiotic relationship, which is a commonplace phenomenon, to endosymbiotic adoption, which is an extremely rare one, or, especially, on the development of the main properties of eukaryotic cells. This question will be examined further in Chapter 17, after we have taken a brief look at a major protagonist in the endosymbiont game: oxygen.

In the meantime, a fundamental difference between the two models under discussion must be underlined. Models based on the fateful-encounter hypothesis all assume that the typical properties of eukaryotic cells were acquired *after* the uptake of the first endosymbionts, which is viewed as the triggering event of the entire transformation from prokaryotic life to its eukaryotic form. In contrast, the proposed primitive phagocyte is seen as having developed *before* the uptake of the first endosymbionts and as having been directly instrumental in this uptake. The importance of this difference can hardly be overemphasized. In particular, eukaryotic proteins are expected to be much older in the former case than in the latter, a point of critical interest in connection with the sequence data discussed above (Hartman and Fedorov, 2002).

16 Oxygen

There is considerable evidence, from the analysis of very old terrains, that Earth's atmosphere at the onset was virtually devoid of oxygen and that its present content of this gas is due to the advanced form of photosynthesis carried out by cyanobacteria and by the chloroplasts of unicellular algae and green plants, a process whereby electrons are extracted from water with the help of energy from light, releasing molecular oxygen. In accordance with this explanation, the amount of oxygen (O_2) in the atmosphere is equivalent to the total amount of carbon (derived from CO_2) immobilized in biomass and in fossil deposits by oxygen-producing photosynthesis (Figure 16.1). This balance, established over hundreds of millions of years, is now being threatened by the excessive return of biocarbon to the atmosphere due to the increasing human consumption of fuel.

An obvious correlate of the geochemical evidence is that the first forms of life were anaerobic and remained so until the time, about 2.2 billion years ago, when the oxygen content of the atmosphere started rising (Bekker et al., 2004). According to a hypothesis publicized by Lynn Margulis under the dramatic name of *oxygen holocaust*, the rise of atmospheric oxygen caused a massive extinction of the life forms that existed at the time. It is, indeed, known that, in living organisms, oxygen readily gives rise to toxic derivatives, including the hydroxyl radical (OH), the

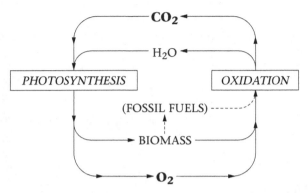

Figure 16.1. *The equivalence between atmospheric oxygen and immobilized organic carbon.* Diagram shows how the amount of oxygen present in the atmosphere is equivalent to the amount of CO_2 carbon immobilized in biomass and, secondarily, in fossil fuels by the mediation of oxygen-producing photosynthesis. Metabolic oxidations and fuel combustion use up oxygen and return an equivalent amount of carbon to atmospheric CO_2. Increased combustion of wood and, especially, fossil fuels is upsetting this balance, leading to a decrease in atmospheric oxygen content (imperceptible under present circumstances) and to a (highly significant) increase in atmospheric CO_2 content (doubling in the last century), with as a consequence a warming of the climate due to the greenhouse effect. Note that the contributions of chemotrophy and of primitive photosynthesis to carbon storage are not included in the diagram.

superoxide ion (O_2^-), and hydrogen peroxide (H_2O_2), all of which can cause considerable harm, as exemplified by today's strict anaerobes – the bacillus of gaseous gangrene is an example – which are killed by oxygen. All aerobic organisms are protected against these effects by specific enzymes, such as superoxide dismutase, peroxidases, and catalase.

According to the oxygen-holocaust scenario, oxygen created one of the most stringent bottlenecks in the history of life, sparing only those organisms that found refuge in an oxygen-free niche and, rare but immensely important, those that happened to develop adequate defense mechanisms in time. From these survivors, it is assumed, a group emerged that progressively turned mere protection into what was to become the most powerful source of biological energy: the use of molecular oxygen as final acceptor in downhill electron transfers coupled to the assembly of ATP.

A popular version of this scenario, to which I subscribed (de Duve, 1995, 2002), sees the primitive phagocyte – or whatever organism became the endosymbionts' host cell – and all the organisms that preceded and accompanied it as having been obligatory anaerobes, which were virtually wiped out by oxygen when it appeared. According to this version, only those forms that acquired oxygen-utilizing organelles managed to squeeze through the bottleneck, allowing eukaryotes to survive and eventually give rise to the most conspicuous and diverse part of the biosphere, including the human species.

This intellectually attractive theory has been shaken in recent years by various hints suggesting that oxygen may not have been as rare on the early Earth as envisaged by the oxygen-holocaust scenario (Lane, 2002). One piece of evidence came from the discovery, by the American J. William Schopf, of 3.5-billion-year-old microfossils that he identified as originating from cyanobacteria (Schopf, 1999). However, this identification, and even the biological origin of the alleged microfossils, have since been contested (Brasier et al., 2002). The controversy is still unresolved, but the cyanobacterial origin of the traces is generally considered doubtful and no longer claimed by Schopf. Not just photosynthesis, however, is considered responsible for the early production of oxygen. It is widely accepted that small amounts of this gas were continually generated by the ultraviolet-induced splitting of water, followed by loss to outer space of the hydrogen released.

The main problem, however, is not whether oxygen was or was not produced, but what level it reached in the atmosphere and, especially, in the waters harboring the living organisms that existed in those days. This depends on the efficiency of the available *oxygen sinks*, which is the name given to systems able to trap oxygen chemically. The main sink was probably ferrous iron (Fe^{2+}), which was present in large amounts in the early oceans and is readily oxidized to ferric iron (Fe^{3+}) by molecular oxygen, resulting in a highly insoluble mixed ferrous–ferric oxide, which eventually gives rise to the mineral magnetite. That this reaction may have taken place on a large scale is supported by the existence of magnetite-rich geological formations – known as banded iron formations because of their stratified structure – dating back to times ranging between 3.5 and 2 billion years ago.

Most workers believe that the iron sink long remained effective enough to protect anaerobic life against oxygen toxicity and that oxygen became threatening only when this sink became saturated, and especially when cyanobacteria started developing in large amounts, releasing increasing quantities of oxygen by photosynthesis. The sink, however, cannot have been effective to the point that no oxygen at all was present. Interestingly, some bacteria adapted to such a situation exist today, combining the characteristic sensitivity of strict anaerobes to substantial oxygen levels with the ability to tolerate the presence of very low levels of the gas, and even benefit from it (Baughn and Malamy, 2004). Called *nanaerobes* (because they grow in nanomolar concentrations of oxygen), these organisms owe their ability to use oxygen to their possession of a cytochrome oxidase presumably characterized, like other enzymes of this type, by a high affinity for oxygen. The molecular phylogeny of this enzyme apparently goes back almost to the LUCA, suggesting that early life may well have been "nanaerobic," rather than strictly anaerobic. This possibility agrees with other findings indicating that biological oxygen utilization may have started very early (Castresana and Saraste, 1995), though this does not necessarily mean that the oxygen holocaust is a myth. Remember that present-day nanaerobes are killed by atmospheric oxygen, just like strict anaerobes.

There is another argument, however, suggesting that the rise in atmospheric oxygen that took place around 2.2 billion years ago may not have had as deadly consequences as was believed. Evidence has been found indicating that the lower part of early oceans may have been kept largely anoxic by an abundance of sulfide-producing bacteria, even when the upper part became progressively enriched with oxygen (Canfield, 1998; Anbar and Knoll, 2002; Knoll, 2003; Arnold et al., 2004). Thus, a luxuriant anaerobic ecosystem could have continued thriving in the depths while aerobic organisms flourished above. There would have been plenty of opportunity, between the two regions, for a graded transition from one form of life to the other.

17 Endosymbionts

As mentioned in Chapter 15, there is considerable disagreement concerning the manner in which the initial relationship between prospective host cells and endosymbionts may have been launched. What followed, however, is fairly well understood, at least in its main lines, from features that are common to the two historically established instances of endosymbiosis. Whether in the case of mitochondria or that of chloroplasts, essentially two kinds of events were involved in the adoption of the organelles.

In both cases, the endosymbionts have lost a major part of their genes, and a certain number of these have been transferred to the host-cell nucleus and integrated within its genome in a manner that allowed replication and transcription of the genes by the local mechanisms. Interestingly, it may happen that such genes or remnants of them remain in the nucleus after the organelles themselves have been lost or become degenerate. This kind of evidence first revealed that the ancestors of certain amitochondriate protists may once have contained mitochondria, a finding that, as we have seen, was brandished against the primitive-phagocyte hypothesis.

The second phenomenon consists in the translocation into the endosymbiont of the protein products of certain endosymbiont genes displaced to the host-cell nucleus. Such proteins are synthesized, like the host-cell proteins, by cytosolic ribosomes instructed by RNA messages transcribed in the nucleus. If the proteins play an essential role in the endosymbiont, their translocation must be operative before the relevant gene is definitively lost by the endosymbiont and transferred to the nucleus.

These mechanisms and the selective advantages that may have driven their development have been discussed in previous works (de Duve, 1995, 2002).[1] They will not be considered further in this chapter, which is devoted to the singularity of the adoption process and its possible causes.

Mitochondria and Hydrogenosomes

Mitochondria are the "power plants" of eukaryotic cells. Surrounded by two membranes, they house in their inner, pleated membrane (cristae) the phosphorylating respiratory chains responsible for the oxidative generation of ATP in the vast majority of eukaryotes. There is clear evidence that all mitochondria are derived from a single ancestral endosymbiont, which is itself derived from ancient bacteria most closely related to present-day α-proteobacteria.

As soon as this fact was suspected, the question arose as to what evolutionary advantage could have favored the observed association between the endosymbionts and their host cells. This question was first answered in the framework of the oxygen-holocaust hypothesis. The assumption was made that the host cell was an obligatory anaerobe that would have been wiped out by oxygen, had it not been rescued from this fate by its oxygen-utilizing endosymbionts.

Arguing against this possibility, however, was the fact that mitochondria, as well as their closest extant prokaryotic relatives, possess the most efficient and perfected protonmotive respiratory chain found in the entire living world. It seemed unlikely that a strictly anaerobic organism could have survived all the time needed by some neighboring prokaryote to acquire such sophisticated systems, waiting, so to speak, in the oxygen-free wings until adoption of the champion oxygen utilizers allowed the

[1] More recent details may be found in Dyall et al. (2004a).

hero to make its spectacular entrance on the aerobic scene. In view of this objection, I suggested, as a more likely explanation, that the rescuers were peroxisomes, which contain oxygen-consuming enzymes of suitably primitive character (de Duve, 1995, 2002).

How this rescue could have happened cannot be pictured at the present juncture, for the origin of peroxisomes is unknown. Whichever way these organelles appeared, it is clear that they could have protected the primitive host cell against oxygen toxicity until the ancestors of mitochondria were adopted as endosymbionts. But if such were the case, what could have been the advantage cells derived from acquiring mitochondria? Energetic efficiency was seen as the most plausible answer to this question. Indeed, if, as had to be assumed, the host cells relied for their ATP supply on primitive systems of the kind involved in substrate-level phosphorylations, acquisition of the highly efficient carrier-level phosphorylation systems present in mitochondria would have resulted in an enormous (almost 20-fold) gain in energy yield, manifestly a decisive selective advantage.

This theory has recently been challenged by findings indicating that mitochondria are most likely related to *hydrogenosomes*. Discovered in my laboratory at Rockefeller University by Miklos Müller (1993), hydrogenosomes are membrane-bounded organelles first identified in the facultative anaerobe *Trichomonas vaginalis*, a parasite of the female genital tract. These organelles are characterized by the ability to support the anaerobic generation of ATP by a process linked to the production of molecular hydrogen; hence their name. Hydrogenosomes can also function aerobically, in which case the hydrogen is diverted oxidatively to form water. Organelles endowed with these properties have been found in a small number of phylogenetically distant protists and also in some fungi, indicating multiple origins.

In recent years, a number of genetic kinships between hydrogenosomes and mitochondria have been uncovered (Martin and Müller, 1998; Müller and Martin, 1999; Dyall et al., 2004a; Sutak et al., 2004), leading to the conclusion that the two kinds of organelles are derived from a common ancestor, which presumably combined a highly efficient oxidative machinery with the ability to survive anaerobically by a hydrogen-generating process. These remarkable findings added a new dimension to the endosymbiosis problem: In addition to aerobic efficiency, the host cell could gain a second evolutionary advantage from its endosymbionts, namely the ability to assemble ATP anaerobically by a mechanism associated with hydrogen production.

A theory based on this second process has been proposed in a landmark paper by Martin and Müller (1998). The theory assumes the joining of an autotrophic, hydrogen-utilizing, archaean prokaryote, possibly similar to present-day methanogens, with a heterotrophic, hydrogen-generating, bacterial prokaryote, in a mutually advantageous association based on the swapping of hydrogen for food. Initially symbiotic, the link between the two partners is assumed to have changed progressively to an endosymbiotic one, with the archaean becoming the host, and the bacterium the guest that became the common ancestor of hydrogenosomes and mitochondria.

Note that the starting premise of this hypothesis rests on a well-known phenomenon. Symbiotic associations between hydrogen-producing and hydrogen-utilizing prokaryotes are found abundantly in nature. What is original to the proposal in question is the alleged conversion from a symbiotic to an endosymbiotic relationship and, especially, the resulting development of an organelle ancestral to mitochondria and hydrogenosomes. The model differs fundamentally in this respect from another model, also based on hydrogen transfer, that was proposed at the same time by Moreira and Lopez-Garcia (1998). In the latter model, which was designed to account for the genetic chimerism of eukaryotes (see Chapter 14), the roles are reversed. Archaean methanogens become the endosymbionts, whereas bacteria (related to δ-proteobacteria, a group different from the mitochondrial ancestors) become the host cells. I shall return to this model at the end of this chapter, when discussing the possible endosymbiotic origin of the eukaryotic nucleus.

The central feature of the Martin–Müller hypothesis, and the reason for its success, is that it traces to a single event the emergence of two key properties of eukaryotic cells: their genetic chimerism, partly archaean and partly bacterial, and their possession of bacterial endosymbionts, taken to be derived from an ancestor combining the properties of mitochondria and hydrogenosomes. According to the hypothesis, the characteristic hydrogenosomal properties were lost in most cases, leaving typical mitochondria (or sometimes only a few of their genes). In a few selected cases, which phylogenies show must have happened independently more than once, mitochondrial properties were lost, leaving hydrogenosomes.

In the lively context, recalled in Chapter 14, of phylogenetic investigations and debates on the chimeric origin of the eukaryotic genome, the hydrogen model has, understandably, attracted a great deal of interest. It does, however, suffer from some weaknesses and would gain by being amended.

First, the transition from symbiosis to endosymbiosis is unexplained by the model. Whereas many examples of symbiosis among prokaryotes are known, only a single case of apparent endosymbiosis has been described (von Dohlen et al., 2001). Cited as proof that "it is possible to establish an endosymbiosis with a non-phagocytotic, prokaryotic host" (Martin and Russell, 2003), this single case is hardly demonstrative compared with the many authenticated cases of endosymbiont uptake by phagocytosis, especially in that the host is not even a free-living prokaryote, but an endosymbiont itself.

Another weakness of the hydrogen model, already mentioned in Chapter 15, is that, like all models based on the fateful-encounter hypothesis, it fails to account satisfactorily for the development of important eukaryotic properties, in particular the cytomembrane system and the cytoskeleton, not counting the nucleus, together with all the key functions associated with these structures. These difficulties are compounded by the fact that the membranes of the alleged host-cell must have contained ether lipids, which were somehow replaced by ester lipids, presumably under the control of genes originating from the endosymbiont (Martin and Russell, 2003).

Finally, the hydrogen model rests exclusively on the endosymbiont's ability to generate hydrogen anaerobically, completely neglecting its aerobic assets, which, surely, cannot have lost their value merely by being associated with a useful anaerobic property. Yet, these assets must have existed at the time the endosymbiotic link was created. It is inconceivable that the hydrogen-generating endosymbiont could later have acquired the perfected phosphorylating machinery that characterizes mitochondria. If, as the model supposes, the endosymbiotic association arose in an anaerobic milieu, it must be assumed that the respiratory machinery remained inactive during the establishment of this association, while staying functionally intact, ready to resume its activity and again prove useful when, as necessarily must have occurred, the system moved to an oxygenated environment. It is questionable whether prolonged preservation of such a complex system could have taken place without any selective pressure.

The existence of aerobic and anaerobic zones separated by a "buffer" interface, mentioned in the preceding chapter, offers a possible loophole. It is conceivable that the endosymbiotic relationship was sealed under strictly anaerobic conditions, as the hypothesis demands, but that the proximity of an oxygenated zone provided opportunities for the respiratory ability of the endosymbiont to prove useful by scavenging infiltrating oxygen (Martin and Müller, 1998).

There are other possibilities. A primitive phagocyte equipped with peroxisomes, as surmised above, would also fit the bill as host cell, provided it derived some advantage from being supplied anaerobically with hydrogen. Also conceivable is a situation where both members of the symbiotic association occur together as endosymbionts in a third cell serving as host. Interestingly, a parasitic protist present in the hindgut of certain cockroaches does indeed harbor endosymbiotic, methane-producing archaeans in intimate contact with hydrogenosomes, which, contrary to other organelles of the group, still contain small amounts of DNA (Akhmanova et al., 1998).

Finally, the possibility that the "hydrogen connection" may turn out to be a red herring and that some other mutual advantage accounted for the sealing of the endosymbiotic relationship between the ancestors of mitochondria and hydrogenosomes and their host cells cannot be excluded at this stage. In fact, the very basis of the new model, which is the common ancestry of mitochondria and hydrogenosomes, still remains open to question (Akhmanova et al., 1998; Dyall et al., 2004a, 2004b). The main problem arises from the fact that, unlike mitochondria, hydrogenosomes seem to have appeared more than once in evolution, with genetic complements that may differ from one lineage to the other (Horner et al., 2000; Hackstein et al., 2001; Voncken et al., 2002). Such traits are not readily reconciled with a single origin from an ancestor that already possessed the typical hydrogenosomal properties (together with those of mitochondria). An alternative possibility, which definitely deserves serious consideration, is that the mitochondrial ancestor initially lacked the properties of hydrogenosomes and later acquired these properties by lateral gene transfer in the course of adaptation to anaerobic conditions, a process that could have happened more than once in distinct lineages. The genes, which may not have been the same in each case, could have originated from another endosymbiont (Dyall et al., 2004a). It is even conceivable that it was this endosymbiont that gave rise to hydrogenosomes and that it acquired from mitochondria the genes it shares with these organelles (Dyall et al., 2004b). Such gene transfers between endosymbionts could have happened directly or by way of the nucleus.[2]

What now of the singularity? According to some defenders of the fateful-encounter hypothesis, a fantastic stroke of luck (mechanism 6) was involved. In the words of Martin and coworkers (Martin et al., 2003, p. 197), "the origin of mitochondria was an unspeakable [sic] rare event." This is a philosophically loaded statement, as it implies that the whole visible biosphere, including our own species, owes its existence to "an unspeakable rare event."

Other possible explanations can be considered, however, especially if the primitive-phagocyte hypothesis is accepted. With such cells, the encounters themselves between potential host cells and guests would have been commonplace, just as they are today. What must have been responsible for the singularity are the events that followed and finally led to the conversion of phagocytized bacteria into endosymbionts. Even if the explanation escapes us, it remains possible that endosymbiotic capture was

---

[2] I have been greatly helped in dealing with this highly specialized topic by extensive correspondence with Patricia Johnson, to whom I am much indebted. The assistance of Johannes Hackstein, who has kindly provided me with much precious information, is also gratefully acknowledged. These colleagues should not, however, be held responsible for any mistakes I may have made in summarizing the published material and in trying to make sense of it.

an indispensable survival condition for the host cells under the prevailing conditions, thereby creating a selective bottleneck (mechanism 2). A bottleneck of the restrictive type (mechanism 3) can also be envisaged, in view of the many conditions that had to be satisfied for a successful integration of the endosymbiont. Remember, however, that the same conditions were fulfilled at least twice (mitochondria and chloroplasts) and cannot, therefore, have been forbiddingly stringent.[3] Finally, there is also the possibility of a "pseudo-bottleneck" (mechanism 4), such that, of the many similar lines that were initiated, only a single one ended up surviving in the common ancestor of all eukaryotes, while all the others fell victim to the natural extinctions that unavoidably trim phylogenetic trees. With no help from fossils and with the genetic lines inextricably intermingled by lateral transfers, paralogies, hidden endosymbiotic inheritances, and other complications, the solution to the problem may never be found; but one can only try. In this connection, the fact that the same problem is raised by chloroplasts, in what appears to be a simpler context, could provide suggestive hints.

Chloroplasts

In principle, the chloroplast situation is more straightforward. These photosynthetic organelles clearly originate from cyanobacterial ancestors that were adopted as endosymbionts by cells that already possessed mitochondria (and peroxisomes). It is generally assumed that these host cells had developed the main properties of eukaryotic cells by the time this event took place, whatever the nature of the host cell that accepted mitochondria, and that uptake of the ancestral cyanobacteria did occur by phagocytosis. There is, thus, no controversy of the kind that exists about mitochondria. As to the selective advantages resulting from the association, they are obvious. Heterotrophic cells that depended on the continual uptake of

---

[3] This argument is not entirely conclusive, for it is possible that cells that have successfully adopted one kind of endosymbiont may more readily adopt a second kind, taking advantage, for instance, of the protein-translocation system developed in the earlier process and simply modifying its targeting signals (Dyall et al., 2004a).

food to survive could, henceforth, rely on their photosynthetic captives to supply them with nutrients, with the sole requirements being light and some mineral elements.

Why, then, did the phenomenon occur only once? Or, at least, why has only a single such phenomenon produced offspring that have survived until this day? The question is a recent one. Choloroplasts were at one time believed to be the products of at least three distinct endosymbiotic events, leading to the separate formation of red, brown, and green algae, of which only the latter evolved further into multicellular plants. The present trend, however, is in favor of a monophyletic origin of all chloroplasts, even though multiple endosymbiotic adoption processes may have taken place, but involving chloroplast-containing cells, or even the products of a previous such event, but not free cyanobacteria (Palmer, 2003).

Using Occam's razor, one is tempted to assume that the singularity has a similar explanation in the two cases, of mitochondria and chloroplasts, and that it is somehow connected with the process of endosymbiotic adoption of a previously free-living prokaryote. Selective optimization comes to mind as an attractive possibility, considering that many different combinations of host-cell and endosymbiont genes may be put to a test in the course of such an adoption process. In such an event, the organelles, rather than being the outcome of improbable chance events, would have arisen as the refined products of natural selection.

Other Endosymbionts

Biologists have long been intrigued by the many symbiotic associations that exist in nature. In the last decades, this interest has extended to all kinds of endosymbiotic associations, involving not only prokaryotes and prokaryote-derived organelles, but also eukaryotic cells and parts of such cells as adopted entities. Endocytobiology has become a distinct discipline, with its special societies, journals, and meetings. Even a brief summary of this fascinating field falls outside the limits of this book. Only a few examples directly related to the origin of eukaryotic cells will be mentioned.

As mentioned above, an endosymbiotic origin of eukaryotic *cilia* and *flagella* has been postulated by Margulis, but this hypothesis is no longer considered likely by a majority of workers.

The origin of *peroxisomes*, on the other hand, remains a tantalizing, unsolved problem. At a meeting organized in 1968, I reviewed the biochemical properties of the various members of the peroxisome family that had been identified at that time in animal and plant cells and in protists, and I came to the conclusion that the common ancestor from which they presumably are derived was a metabolically rich entity that could possibly have originated endosymbiotically from some primitive aerobic prokaryote but had kept no trace of this origin, having lost all of its conserved genes to the nucleus of its host cell (de Duve, 1969). As far as I know, this possibility remains open but lacks empirical support.

It is to be hoped that future phylogenetic studies will help to clarify this problem. Peroxisomes are important organelles present in the great majority of eukaryotic cells. Their hydrogen-peroxide-centered oxidative machineries suggest that these organelles may have played an important role in protecting early eukaryotes against the rising level of atmospheric oxygen. No model purporting to account for the origin of eukaryotic cells can be complete without providing an explanation for the origin and evolutionary function of these intriguing organelles.

Finally, attention must be given to the possibility that the eukaryotic *nucleus* may originate from an ancestral endosymbiont. This possibility, which is one of the explanations proposed to account for the genetic chimerism of eukaryotic cells (see Chapter 15), should not be confused with the well-established fact that endosymbionts have abandoned a large fraction of their genes to the host-cell nucleus, with as a result a certain degree of fusion between the two genomes. What is referred to here is the theory, advocated notably by Hartman (1984; Hartman and Fedorov, 2002), according to which the eukaryotic nucleus belonged to an archaean prokaryote that was endosymbiotically adopted by a hypothetical phagocytic host cell, or "chronocyte," with an RNA genome. How the DNA-based genetic system of the endosymbiont could have become integrated within an RNA-based economy is far from clear, however, and is not explained by the author.

Also to be mentioned in the same context is the "syntrophic" model, cited above, of Moreira and Lopez-Garcia (1998), according to which envelopment of a hydrogen-utilizing, archaean ancestor by hydrogen-producing bacteria led to an endosymbiotic relationship in which the archaean would have given rise to the nucleus, and the bacteria to the cytoplasm. Except for a vague allusion to the development of an endomembrane system from the peripheral membranes of the enveloping bacteria, the authors of this model hardly discuss the question, although crucial, of the manner in which the partners of the putative association might have evolved to concentrate the informational genes of one and the operational genes of the other in a single nucleus derived from the archaean endosymbiont (see Chapter 14).

18 Multicellulars

The history of multicellular organisms represents the latest and, thanks to the fossil record, best-known stage in the long saga of life on Earth. It has been the subject of many learned treatises and discussions, which can profitably be consulted by any reader interested in the details. My purpose, in this chapter, is to focus solely on my chosen topic, namely, that of singularities, an aspect that, surprisingly, has attracted relatively little attention from the experts. Yet, singularities are a prominent feature of the trails that have led to present-day multicellular organisms. As far as is known, all land plants, all fungi, and all animals are derived each from a single common ancestor. Each new major group that appeared later in the evolution of life seems to go back to a single stem organism.

The Founders

Exactly when eukaryotic cells first formed multicellular associations remains uncertain, largely because many such associations may have existed without leaving any recognizable fossil trace. It seems likely, however, that single-celled protists thrived and evolved on Earth for as long as several hundred million years before giving rise to the multicellular organisms whose descendants surround us today. Why did it take so long?

A plausible answer to this question is that we are not dealing here with associations as generally understood, that is, entities composed of two or more genetically different cell populations that came together under the pressure of natural selection and are held together by the mutual benefits they derive from joining forces. Many such symbiotic colonies exist, even in the prokaryotic world. True multicellular organisms originate from a *single* cell – either haploid or, more often, diploid and arising from the fertilization of one haploid cell by another – that produces, in the course of successive divisions, distinctly differentiated cell types that all possess the same genome but express different parts of it. The key to the development of such organisms lies in the ability to regulate *transcription* in such a way that different sets of genes are silenced or activated in different descendants of the mother cell. Furthermore, these phenomena need themselves to be coordinated in time and space so as to give rise to sets of differentiated cells harmoniously organized into tissues, organs, systems, and, in the end, whole organisms.

An explanation that could conceivably account both for the late appearance of these multicellular organisms and for their singular origin is that their formation was the outcome of extremely improbable events that required some fantastic stroke of luck to occur (mechanism 6). Arguing against such an explanation is the fact that multicellularity based on cellular differentiation was developed independently on at least three separate occasions: by green algae in the case of land plants,[1] by some primitive yeast cells in the case of fungi, and by some motile protists (choanoflagellates) in the case of animals.

In addition, there seems to be no need of special circumstances to induce sister cells to stick together after their birth. They could do so because they shared a common housing, as do the cells of plants and fungi. Or they could bind to each other by way of cell-adhesion molecules, as is the case with animal cells. Thus, their "choice" between remaining together and gaining freedom would depend on the relative advantages of the two states, as explored by various mutational events. The myxomycete (slime

---

[1] Marine red and brown algae are themselves believed to be the products of separate events.

mold) *Dictyostelium* offers an interesting example of an environmentally regulated alternation between the two states. In times of plenty, the cells behave like free, amoeboid protists successfully pursuing their individual existences. When stressed (by scarcity of food, for example), the cells congregate into a fruiting body that, sometimes through sexual reproduction, gives rise to spores, which serve to preserve the species until the advent of better days. Revealingly, the distress signal calling for congregation is sent by means of secreted cyclic AMP, a major ATP-derived molecule with multiple regulatory functions, which, in the present case, acts by turning on genes for cell-adhesion proteins. One readily imagines how certain mutations of transcription-controlling genes involved in cellular differentiation may provide a selective advantage to a given cell association, which thereby becomes stabilized.

For the above reasons, one is tempted to suspect that the initiation of a successful association among two and, later, more kinds of differentiated offshoots of the same mother cell could be a fairly commonplace event, which may have happened more than once in each kingdom. If so, where do the singularities observed in the present-day world come from?

To begin with, we must remember that we are dealing here with events that took place at least 600 million years ago, leaving as only traces (other than rare fossil remains) the end products of the lineages that have survived for all those very long times to give us an inkling of what their forebears may have been. Many other similar branches could have started and become extinct as a result of a variety of unrelated chance factors. In this case, we would be dealing with what I have called a pseudo-bottleneck (mechanism 4), not an unlikely explanation, especially in that the present living world clearly does not include representatives of all evolutionary intermediates that have existed. What it provides is a spotty sampling, consisting of lines that, for one reason or another, escaped extinction and extend to the present.

The possibility, however, that some real bottleneck could account for the singularities cannot be excluded. It is certainly conceivable that, as in many earlier instances considered in this book, evolving life had the opportunity to explore a vast array of possibilities and ended up, at each stage, with an optimal or near-optimal combination of genes, given the circumstances (selective bottleneck, mechanism 2). Such a process could have been further narrowed down by internal restrictions (restrictive bottleneck, mechanism 3). As will be seen, such restrictions probably became increasingly stringent in the course of evolution, as genomes became more complex.

The Stem Organisms

The tree of life has grown, much like real trees, by a combination of branch lengthenings and bifurcations. These events are shaped by reproduction, as channeled by genetic changes and environmental conditions. In prokaryotes and single-celled eukaryotes that multiply by binary division, events are relatively simple. Branch lengthening reproduces existing genomes either unchanged or in a form progressively modified by mutations, as screened by natural selection. Forks occur when a mutant cell detaches from the main line and starts a new branch, often adapted to a different environment.

Things are more complicated with sexually reproducing organisms. Reproduction does not operate with entire genomes; it continually shuffles existing genes into new combinations produced in the course of meiosis and fertilization, leaving it to natural selection to weed out the less advantageous combinations. Modified genes are introduced into this pool by mutant germ cells that happen to participate in a successful fertilization event. Forks are started when genetic variants separate from the parent population and inaugurate a new line, usually, but not always, thanks to isolation in a different environment.

An examination of phylogenetic reconstructions discloses two different kinds of forks. In one, existing body plans suffer only minor modifications, often associated with greater evolutionary success in a given environment, but without significant long-term impact on the evolution of life. In this form of evolution, which textbooks generally call microevolution and I like to call *horizontal*, there is merely creation of diversity within existing groups. More than a million insect species probably exist, from ants and cockroaches to fireflies and praying mantises; but all are insects, descendants of a single ancestral insect form. Similarly, cods, stingrays, sea horses, and barracudas all have arisen by horizontal evolution from a single ancestral fish species.

A much rarer kind of fork is created when the responsible genetic change initiates a series of successive alterations leading to a profoundly reorganized body plan. Thus it happened, for example, that some ancient fish species, instead of merely evolving horizontally to produce other fish, suffered a change that opened the way toward the first primitive amphibian, through a long succession of stages in which a number of traits were acquired that allowed the organisms thus endowed to pursue their existence out of water. Some amphibian species likewise led to the first reptile, a reptile to the first mammal, and so on. Called macroevolution in textbooks, and *vertical* evolution in my terminology, this process joins the founding organisms (stem organisms) of all major phyla by a continuous line (the trunk of the tree or, better said, its animal master branch), from which side branches created by horizontal evolution spread out into an increasing number of ramifications. The evolution of plants and fungi shows a similar structure.

Lest this image be misunderstood, it must be noted that master branches are recognized only by hindsight; they may include many intermediates that would have been seen as the products of horizontal evolution at the time they appeared. For example, lungfish, which are found, notably, in certain African lakes, look to us like typical products of horizontal evolution adapted to conditions of alternate wetness and dryness. Only our knowledge of present-day animals allows us to identify lungfish as probably related to the ancestors of amphibians. A more correct way of looking at these events, therefore, is to see horizontal evolution as branching from the stem organism in all sorts of different directions and ending almost invariably with specialized forms, well adapted to their environment but of little evolutionary potential, by all appearances evolutionary dead ends. On exceptional occasions, the product of this process is revealed by subsequent history as having been a stem organism from which a completely new set of evolutionary pathways was initiated. From horizontal, the process has become vertical.

This consideration may seem to suggest that there is no basic difference between the two kinds of evolution in terms of mechanisms, but
only in terms of outcome. Compare, for example, the pathways that led
from the ancestral fish to the first sea horse, on one hand, and to the stem
amphibian, on the other. In both cases, we may assume the occurrence of
a string of successive mutations defining a set of increasingly specialized
traits and ending with two highly specialized forms. The mechanisms are
basically the same. Some difference, however, must account for the fact
that the evolutionary potentials of the two products are very different,
endowing them with entirely different futures.

This difference concerns primarily the nature of the genes that undergo the critical mutations. The modifications that initiate the rare
branches destined to evolve vertically have as targets master genes –
homeotic genes, for example – that control the pattern of development
of an organism. One clearly identified such change is correlated with an
increase in the number of distinct, differentiated cell types, a typical parameter of vertical evolution in both the plant and the animal kingdom.
Starting with two, this number has grown steadily with evolution, reaching several tens in the most advanced plants, and more than two hundred
in the higher animals.

A feature common to horizontal and vertical evolution is what may be
called *funneling*, a progressive narrowing of the range of directions open
to further evolution. Whether the end product was a sea horse or a primitive amphibian, it is clear that evolution became increasingly restricted –
and committed – as it proceeded. These constraints are both externally
and internally imposed. In many cases, adaptation to a given kind of
environment tends to limit the ability of organisms to survive in other
environments. In addition, the structures of genomes and the peculiarities of genetic mechanisms may contribute further to the funneling of
evolutionary processes. Thus, genes are known to vary greatly in their mu-
tability, to the point that certain hypermutable regions exist in genomes,
presumably selected by virtue of special advantages (Caporale, 2003). Furthermore, as already mentioned, certain genes, such as homeotic genes
and other genes coding for transcription factors, exert a much more important control over development than others and are thus privileged

targets for mutations influencing evolution. As knowledge of genome organization increases, many other such factors, of too specialized nature to be mentioned here, are coming to light.

In conclusion to this cursory survey, we must take a look at the evolutionary development of the most remarkable of all multicellular organisms, modern humans.

19 Homo

Highlighted by a few discoveries that have inflamed the imagination of the greater public, the history of our origins has become, thanks to modern means, an object of intensive investigations and discussions, related in numerous recent works. Little of this will be covered in the present chapter, which, like the preceding one, will be limited to data directly relevant to the main topic of this book.

A Bird's-Eye View of Human Evolution

Figure 19.1 brings together some illuminating clues to the ancient, still poorly understood pathway whereby our species has evolved from the last ancestor we have in common with chimpanzees, our closest extant relatives. In this graph, some of the better-known hominids are represented, on one hand, by the approximate time range of their existence and, on the other, by their estimated brain size, the most dramatically changing parameter of this crucial development. Several features of this graph deserve to be noted.

First, there are two landmark singularities. The first goes back to the time, between five and seven million years ago, when the human line separated from that leading to chimpanzees. According to currently available

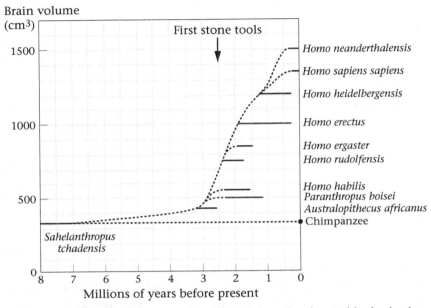

Figure 19.1. *Human evolution.* Horizontal bars show the observed brain size (ordinate) and time of existence (abscissa) of a number of hominid and prehominid species (data from Carroll, 2003). Broken curve represents the possible course of brain expansion, as estimated from the data.

evidence, all known hominid branches seem to be offshoots of this single line. The second singularity, which is solidly established by molecular data, indicates the origin, roughly 200 000 years ago, of all modern humans worldwide from a single ancestral group, often reduced in the popular press to a single couple some commentators do not hesitate to identify with the biblical Adam and Eve.

This interpretation stems from a misapprehension. It is true that all present-day humans have been traced to a single female ancestor and to a single male ancestor. These identifications rest on the comparative sequencing of two special kinds of DNA: on the female side, mitochondrial DNA, which is transmitted exclusively through the maternal line (only oocytes contribute mitochondria to the fertilized egg); and, on the male side, genes associated with the Y chromosome, which is transmitted from father to son.

These facts do not, however, mean that the two ancestors formed a couple. They belonged to an interbreeding population estimated to have contained up to several thousand individuals of each sex. As time went by and generation followed upon generation, it necessarily happened, for all sorts of reasons, that certain females failed to produce daughters, with as a consequence that their mitochondrial DNA was not perpetuated. The Y genes of males that produced no sons suffered the same fate. Eventually, after a time that theory has shown to be equivalent to twice the generation time multiplied by the number of same-sex members of the group,[1] only a single line will have escaped this phenomenon of progressive random attrition. Thus, we have here a perfect example of a singularity arising from a pseudo-bottleneck (mechanism 4), which, however, applies only to certain genes, not to a species or phylum.

Also notable in Figure 19.1 are the great widths – of up to considerably more than one million years – and extensive overlaps of the bars representing the times during which the species are known from the fossil record to have existed. These features indicate that the species shown are not true intermediates in the prechimpanzee-to-human pathway, but offshoots of such intermediates; they belong to side branches that detached from the main branch and eventually became extinct. Their traits, therefore, do not necessarily reflect those of the intermediates, although some features, for example, brain size, were most likely inherited rather than developed in the course of evolution. The most recent such side branch is represented by the Neanderthals, which disappeared a mere 35 000 years ago and coexisted for a while with authentic humans, often designated Cro-Magnons, that were present at the time.[2]

---

[1] To give an example, a population that lived 200 000 years ago, with a generation time of 20 years, could have contained up to 5000 individuals of the same sex and end up today including descendants of only a single one. This number could, of course, have been smaller. The critical value is the time when all lineages but one had become extinct.

[2] This book was already written when the startling news came of the finding (Brown et al., 2004; Mirazon Lahr and Foley, 2004; Morwood et al., 2004), on the Indonesian island of Flores, of fossil bones, no more than 18 000 years old, displaying obviously hominid characters, and surrounded with primitive stone artifacts and animal remains indicative of hunting and butchering. The surprising aspect of the finding is that the individual in question, presumably a female, was only about 1 m tall and had a cranial capacity of only 380 cm[3], which would make her smaller than the celebrated australopithecine Lucy,

A rough guess at what the actual pathway may have been is represented in the graph by a broken line, which has been drawn on the assumption that the earliest detected trace of a given species lies close to the point where its branch detached from the main branch. Although grossly oversimplified, the resulting curve is very revealing, showing first that the brain size increased only moderately during the first three million years of hominization, approximately up to the time of the famous Lucy and of the intriguing beings whose footprints were discovered by Mary Leakey in the petrified ashes of Laetoli, in Northern Tanzania. Many workers believe that the most important traits acquired during that time were related to bipedal walking – evidenced by the Laetoli footprints – and to the corresponding greater use of hands for prehension and other functions, possibly favored by natural selection in response to a change in habitat from forest to savannah.

Then, about three million years ago, there occurred a period of rapid brain expansion that led, in some two million years, to an almost threefold increase in brain volume. This astonishing phenomenon, which is explainable only in small part by a larger body size, coincided with the manufacture of increasingly perfected stone tools. It is difficult not to see a causal correlation between the two phenomena, possibly related to a mutually reinforcing interchange between brain and hands leading, on one side, to greater dexterity and, on the other, to a better ability to design shapes for a purpose and to transmit acquired skills. Some of the key intellectual traits that characterize humans may have begun to develop during that period. But it must be remembered that the individuals concerned were not yet truly human, a point sometimes forgotten in references to the Stone Age.

Finally, after a period of slowing down, a new, more modest upsurge in brain size seems to have happened, associated with the blossoming

which is about three million years old. The authors of this astonishing discovery take as the most probable hypothesis that this species, to which they have given the name of *Homo floresiensis*, may originate from representatives of *Homo erectus* that settled on the island as early as 800 000 years ago and that it evolved toward dwarfism (a phenomenon known to occur on isolated islands), surviving for a very long time after the parent *erectus* population had become extinct.

of cultural evolution and, possibly, with the first appearance of an articulated language.[3] It may seem surprising, in this connection, that the Neanderthals, who reached a distinctly lower cultural level than their contemporary Cro-Magnons and may even, according to one theory, have lacked the capacity for speech, actually had bigger brains than modern humans. They also had bigger bodies, however, which could account for that. In addition, brain size, even corrected for body size, is only a rough index of mental ability. What counts is the wiring of the brain and, especially, the degree of development of the cerebral cortex.

The case of the Neanderthals raises the problem of the possible contribution of the more primitive species to the making of *Homo sapiens*. It is known that, some time between two and one million years ago, populations of *Homo erectus* started moving out of Africa, where they originated, and migrating into vast regions of Europe and Asia, where their fossil remains have been uncovered.[4] According to one theory, known as the *multiregional* model, these populations evolved locally and independently into modern humans. A watered-down version of the theory allows for the possibility that the populations were helped in this evolution by breeding with more recent invaders. In opposition to the multiregional model, the single-origin, or *out-of-Africa*, theory holds that modern humans arose from a single root, roughly 200 000 years ago, in Africa, whence they radiated into all parts of the world, while all the other populations, whether in Africa or elsewhere, became extinct.

This controversy, which was quite acrimonious at one time, has now been settled to majority satisfaction (Lewin, 1996). We have seen above that molecular data unambiguously support the single-origin theory. Even Neanderthals, which coexisted with Cro-Magnons for a while, raising the possibility of interbreeding between the two, have apparently not contributed significantly to recent human ancestry, as indicated by the sequences of mitochondrial DNA extracted from fossil Neanderthal bones and, more recently, from remnants of their contemporary Cro-Magnons

---

[3] The inflection shown could very well be an artifact due to the device followed to draw the curve. Make the appearance of *Homo heidelbergensis* earlier than shown, and the curve would flatten out progressively without showing a secondary break.

[4] See footnote 2 above.

(Caramelli et al., 2003). These conclusions have been lately confirmed by the discovery, in Ethiopia, of 154 000–160 000-year-old *Homo sapiens* fossils clearly ancestral to modern humans and distinctly different from Neanderthal precursors of the same age (White et al., 2003). It thus appears that, as shown in the graph, the two groups separated several hundred thousand years ago and probably never interbred.

Mechanisms

The most striking feature of Figure 19.1 is the truly staggering rate – in terms of evolutionary kinetics – at which the brain has expanded during the rapid phase of hominization, tripling in a mere two million years the size it took some 300 times as long to reach in the course of animal evolution: an almost thousandfold acceleration. Note, however, that this still amounts to a very small increase in absolute terms: little more than 10 mm^3 – a mere pin's head – per generation at peak rate. Note also the painstakingly slow rate of the process in the framework of human history. Hundreds of millennia, that is, many times the entire historical record, were needed just to fashion a roughly hewn stone into a thin blade, a hook, or an arrowhead. It is clear that the primitive tools did not, like ours, evolve by way of cultural evolution; they bear witness to a slow biological evolution linked with brain expansion.

Also striking is the sigmoid shape of the main part of the curve, which is typical of an autocatalytic process constrained by some upper limit. The early autocatalytic character points to a mechanism in which each increment in brain size increased the probability of a further increment. Two possibilities come to mind to explain such a process. The bottleneck could be selective (mechanism 2) and caused by the growing demands, to be met only with a better brain, of a challenging environment from which there was no turning back; or it could be restrictive (mechanism 3) and created by increasingly stringent inner constraints. Both factors could, of course, have been involved at the same time to various degrees.

Whatever external and internal factors were responsible for the bottleneck, an additional fact of central importance needs to be addressed, namely the actual occurrence, at each step, of the genetic changes needed for further advance. Incredible luck (mechanism 6) is often invoked to account for this fact, with as implicit or explicit corollary the great improbability of humankind's ever arising. Figure 19.1 belies such an interpretation, showing that not just one but a large number of successive steps were involved in brain expansion. As pointed out in the Introduction, a single fantastic stroke of luck is perfectly plausible, but not a succession of the same such phenomena. We are forced to admit that the range of possible mutations the system was allowed to explore at each step was extensive enough to include the required genetic change with a high level of probability. This is a surprising conclusion, considering the small numbers of individuals and of generations involved; it strengthens the case in favor of inner constraints, suggesting that the factor responsible for the exponential feature of the curve lay in the structure of the genome itself, with each brain-expanding mutation increasing the probability of occurrence of a further mutation of a similar kind.

As already mentioned, the later progressive flattening of the curve indicates that the runaway process of brain expansion was progressively braked by some limiting factor. It is tempting to assume that this limit was imposed by the growing incompatibility of the expanding skull with the dimensions of the female pelvis. Much has been written on the difficulties of human childbirth and on the premature state of human newborns as compared to many animals. A widely accepted theory (neoteny) attributes those traits to natural selection, which is taken to have favored them as an acceptable trade-off for a bigger and, presumably, better brain. If this explanation is correct, then the second upsurge in brain size, if authentic,[5] that led to Neanderthals and to modern humans (see Figure 19.1) could signal some key event in the acquisition of neoteny.

There is one major difficulty with the proposed interpretation: Why only once? If hominization was so stringently channeled, why does none of the side branches manifest any tendency to evolve similarly? We have seen that this possibility, defended by the holders of the multiregional theory, appears to be ruled out by currently available evidence. This is all the more surprising in that some of the branches have extended over huge

---

[5] See footnote 3 above.

spans of time (up to 1.7 million years in the case of *H. erectus*), apparently without ever undergoing the change that would have propelled them in the direction of a bigger brain.

No simple answer to this question comes to mind, except that it may be said to apply to almost any instance of horizontal evolution. Perhaps side branches never encountered the challenges that would have made a better-performing brain an almost indispensable asset. After all, chimpanzees and gorillas still make do with their limited mental endowment in their natural habitat, although they may not do so much longer in the face of human menaces. Another possibility is that side branches evolved in a way that rapidly closed the genetic door to further brain expansion. The problem, it should be noted, is not basically different from those we have met in the preceding chapter in discussing other singularities. Why, for example, did not other lungfish evolve independently to give rise to amphibians? The mere fact that we are dealing with humans does not necessarily alter the basic elements of an evolutionary problem.

Where Are We Going?

The above discussion raises the interesting question of the future of the human species (de Duve, 2002). Clearly, there is no conceivable chance that history could repeat itself. With more than six billion individuals filling every nook and cranny of our planet, there is no possible prospect of some *super-sapiens* group somewhere prolonging vertically the curve of Figure 19.1, leaving *sapiens* to extend horizontally and drift into extinction like the other side branches.

Neither is it imaginable that the whole of humankind could ever evolve better brains under the sole aegis of natural selection acting on chance mutations. Even if our brain should be perfectible – which it most likely is – our present societies could not provide an opportunity that would allow natural selection to favor the relevant genetic changes. Suppose, for example, that a promising combination had been present in the genome of Moses, Michelangelo, Beethoven, Darwin, Einstein, or anyone else, for that matter; there could have been no way for this combination to be propagated on the strength of some selective advantage. On the contrary, our societies favor the opposite trend, if anything.

Remaining to be considered, however, is the likelihood that human intervention will, if not replace natural selection, at least contribute increasingly to orient it. Humans have been modifying their environment ever since they inaugurated sedentariness, agriculture, animal breeding, and, more recently, medicine and industry. Human impacts on the environment may even have started earlier if, as many anthropologists believe, hunting was responsible for the disappearance of such prehistoric mammals as mammoths, aurochs, and saber-toothed tigers. Modern biology has enormously increased this power, opening the possibility that humans will more and more take their fate in their own hands and use the knowledge they have acquired and the technologies derived from this knowledge to direct biological evolution, including their own.

Despite the safeguards and prohibitions that are being erected in many parts of the world to stem the development of the new biotechnologies and, especially, their human applications, one does not see how the movement that has been launched could ever be completely brought to a standstill. There will always be a place, somewhere, where later generations will be pursuing the shaping of our genetic future, using tools of ever greater efficacy, reliability, and safety. But will they make the right choices? That is far from certain. The decisions involved will require much collective wisdom, a trait natural selection does not seem to have favored to the same extent as it has intelligence, creating a disparity that may well prove fatal to the human species.

Indeed, the eventuality that our line will finish like all the others on Figure 19.1, but with no hopeful offshoot emerging on top, deserves seriously to be entertained. In biological evolution, extinction is the rule, rather than the exception. The brain that has made our success could also be our undoing, simply not being good enough to manage its own creations. Should this happen, perhaps enough may be left of the ruins for evolution to resume its upward course under more favorable conditions.

Even if *Homo sapiens* should be irretrievably lost, together with all their technologies and civilizations, there would still remain plenty of time for the adventure to start all over again, possibly with a happier ending.

According to cosmologists, Earth should stay fit to harbor life for at least 1.5 billion years, possibly up to 5 billion years, when the expansion of the sun will render the planet definitively uninhabitable. There is thus ample time for another journey toward a better brain to be inaugurated five, fifty, or five hundred million years from now, long after the demise of *Homo sapiens*. Such a journey could start from some primate or, even, from a lower animal branch, eventually rising above the present human level thanks to a more harmonious combination of genes favoring intelligence and wisdom. Such a prospect is, of course, of little solace to present-day humans. But it could serve as a warning. One can only hope that our descendants will, preferably under the guidance of reason but, if not, under the pressure of natural selection, learn to make better use of their brains than their forebears have done.

20 Evolution

Except for the creationist fringe – regrettably more than a fringe in certain parts of the world – that gives more credence to words written some three thousand years ago than to fossilized vestiges of ancient life, isotope-decay measurements, or molecular sequences, the *fact* of biological evolution is accepted by all those acquainted with the evidence and by the many educated people who, without being specialists themselves, value the scientific approach and are ready to trust its conclusions, especially when these are unanimously endorsed by those who are competent to appreciate the underlying evidence.

The *mechanisms* of evolution, on the other hand, still form the object of debates among experts, but these debates deal more with details and, even, semantics than with the general process of evolution. The main principles, first enunciated by Darwin, of continuity by heredity, variation by genetic modifications, competition among variants for available resources, and screening by natural selection according to the organisms' ability to survive and reproduce under prevailing environmental conditions, are broadly subscribed to. Furthermore, the essentially accidental nature of the genetic modifications responsible for the variety and their lack of intentionality or foresight are seen as conclusively established by modern molecular biology. Only a small minority of scientists dispute the validity of the neo-Darwinian theory, accepting it only for horizontal evolution, but not for vertical evolution, which they claim, against the opinion of most biologists, has, in certain key events, been guided by some finalistic principle of unknown nature, the agent of *intelligent* design.[1]

What remains the main bone of contention is the *probability* of evolutionary events, still the object of heated controversies, which engage not only the experts themselves but a vast number of representatives of other scientific disciplines, theoreticians of all kinds, teachers, social scientists, politicians, jurists, ethicists, philosophers, and even theologians, all interested in the implications of our understanding of biological evolution for their respective fields or occupations.

Chance or Necessity?

It is fair to say that the dominant doctrine in relation to this question is what I have called the *gospel of contingency*, which owes its credit both to the apparently impeccable logic of its foundations and to the forcefulness and fame of some of its proponents, emblematic figures such as Jacques Monod and François Jacob in France, and Stephen Jay Gould in the United States. The argument is straightforward: the course of evolution is demonstrably determined by chance mutations that happen, again by chance, to occur in an environment propitious to the survival and reproduction of the mutant organisms. Chance multiplied by chance – chance squared, so to speak – at the origin of each of the millions of forkings that have traced the tree of life: The inference seems inescapable that the course of evolution has followed unpredictable, nonreproducible pathways ruled by contingency. "Rerun the tape," as unforgettably imaged by Gould, and the outcome is bound to be entirely different.

With all the diffidence required from an onlooker, interested but not a member of the inner circle, I wish to draw attention to a possible flaw in this argument: evolutionary events do not have to be improbable just

---

[1] A detailed discussion of the intelligent-design hypothesis would not fit within the framework of this book. Readers wishing to know more on the topic will profitably read the critical work devoted to it by Kenneth Miller (1999), a biochemist who owns, in addition, the interesting peculiarity of declaring himself a practicing Catholic. See also my discussion of sequence spaces, Chapter 8.

because they depend on chance. To my knowledge, this point, although self-evident, has not received the attention it deserves. Most often, the tendency among biologists has been to equate fortuitousness with unlikelihood, if not explicitly, at least implicitly. As I have argued, here and elsewhere (de Duve, 1995, 2002), this reasoning mistakenly confuses the notion of randomness with that of unbounded possibility. That is clearly wrong. Chance always plays within certain limits, be they the number of faces on a coin or die, the number of slots in a roulette wheel, the number of tickets sold by a lottery, or, for that matter, the number of possible mutations.

As already stated in the Introduction, chance does not exclude inevitability or, more precisely put, near-inevitability, as the door must always be left open for the odd exception. What counts is the number of opportunities provided for a given event to happen, relative to the event's probability. As illustrated by the calculations of Table I.1 in the Introduction, occurrence of a chance event may be ensured with a 99.9% probability if it is provided with a number of opportunities to occur roughly equal to seven times the reciprocal of its probability. Even a seven-digit lottery number, with one chance in ten million of coming out, may be almost bound to win if 69 million drawings are performed.

The lottery example also illustrates the limitations of the argument. That chance does not exclude inevitability may be true in theory. In practice, however, it is true only to the extent that the required number of opportunities is realistically possible. Thus, the contingency theory could still be correct because of simple physical restrictions. Should the number of possible mutations in a given situation exceed by far the number of mutations that can happen in reality, considering the numbers of individuals involved and the times available, fortuitousness clearly will be in the driver's seat. An intuitive perception that such may often be the case probably lies behind the contingency doctrine.

Calculations have shown that this restriction does not apply to point mutations (replacements of one base by another), due to replication errors. As indicated in Table I.1, it takes some 20 billion cell divisions for a given such mutation to occur with a 99.9% probability. This number may seem enormous but, in fact, amounts to the number of stem-cell divisions that occur in our bone marrow in about two hours in the course of red-blood-cell renewal. Admittedly, point mutations probably play a minor role in evolution. It could be that truly important mutations may have lower probabilities of taking place. Even so, given the large number of individuals involved in evolution and the long durations available, enough opportunities may often be provided to compensate for such low probabilities. In many evolutionary situations, extensive, if not exhaustive, exploration of the mutational space may well have taken place, with selective optimization (given the conditions) as a consequence, just as may have happened with the RNA and protein sequence spaces (see Chapters 7 and 8).

This contention might look purely subjective and gratuitous, or even inspired by some preconceived idea,[2] were it not supported by a number of observations. One is drug resistance, a major hazard now that antibiotics, chemotherapeutic agents, and pesticides of various kinds have been introduced. In less than half a century, a large number of pathogenic microbes, either prokaryotic or eukaryotic, as well as many undesirable plants, fungi, and animals, have become resistant to the chemicals developed to eradicate them. Cancer cells can become so in a matter of months. It is clear that the genetic traits responsible for resistance did not develop in response to the drugs – except by chance, if the drugs happened to be mutagenic – and still less for the purpose of evading the drugs. The traits were part of the range offered by the mutational field, and they emerged by natural selection, following introduction of the drugs into the environment.

Mimicry is another example. Everyone is acquainted with animals that look so much like their surroundings that they become almost invisible and thereby elude attack by predators. Obviously, those traits did not develop for the purpose of protective covering. Neither were they copied. Mechanisms for such a phenomenon do not exist. Like drug resistance, mimicry traits were part of the mutational field and emerged by natural selection. The same may be said of biodiversity in general. Often presented as evidence of contingency, the huge diversity of living forms

---

[2] Critics have not failed to accuse me of ideological bias, but without addressing the scientific reasoning supporting my position (Szathmary, 2002; Lazcano, 2003).

can also be seen as illustrating the richness of the mutational field. So does the frequent observation that random mutagenesis often suffices to bring about a specifically defined goal. An example, which I have mentioned before (de Duve, 2002, p. 108), is the 20-fold improvement in penicillin yield that was achieved, shortly after the Second World War, by the simple device of "carpet-bombing" cultures of *Penicillium notatum* with X-rays.

The conclusion I have drawn from these various facts is that mutations were rarely a limiting factor of evolution. There may be exceptions, of course. But, on the whole, it seems likely that, more often than not, evolutionary "choices" were made by the environment, which caused natural selection to single out certain specific genetic traits among a wide array of possibilities (selective bottleneck, mechanism 2). The most extreme instance of such situations is selective optimization, which obtains when evolving systems have the opportunity to subject to the screening effect of natural selection virtually the entire set of possible genetic changes, so that optimal adaptation to environmental conditions is achieved. Telling examples of such a process have already been encountered in our account of the early events in the origin of life. The exquisite adaptation of many living organisms to their surroundings could well have arisen by a similar process.

Another important factor to be taken into account is the phenomenon of evolutionary funneling caused by the many external and internal constraints that narrow down the mutational field as evolution in a certain direction proceeds. Many evolutionists have commented on what appears a progressive acceleration of evolutionary processes once some sort of commitment has been made. The expansion of the human brain, discussed in the preceding chapter, is a typical example of such a phenomenon.

It follows from all of the above considerations that evolutionary pathways may often have been close to obligatory, given certain environmental conditions, rather than contingent and unrepeatable, as received wisdom contends. The growing number of instances of convergent evolution that are being uncovered illustrate this point in a particularly impressive way. Anteaters, felines dependent on hunting, and herbivores built for speed, just to give a few examples, have independently developed essentially the same kind of specializations in widely different parts of the world (Conway Morris, 1998, 2003). So have underground mammals

(Nevo, 1999). Apparently, there are not so many solutions to the problem of, for instance, subsisting on ants or living underground. More remarkably, under sufficient selective pressure, the same solution may be found time and again. The main defender of this thesis, Simon Conway Morris, has rested it on a solidly documented and, by all appearances, objective and unbiased analysis (Conway Morris, 2003), which, however, would have been more convincing if the author had not at the same time taken on the origin of life the almost opposite stand that it may have been a highly improbable event, possibly unique in the entire universe, not failing to link the two theories into a human-centered, theological conception (see footnote 3 in Chapter 12).

The Environment in Charge?

If the views outlined above are correct, the main role in evolution often belongs to the environment, which actually makes the decision, by way of natural selection, among the wide array of choices offered by the mutational field. The examples mentioned above make this clear. Without drugs, pathogens would not develop drug resistance. There would be no stick insects without branches, no leaflike mantises without green leaves, no gray soles without a sandy sea bottom, and so on. In fact, much of biodiversity may be seen as resulting from responses of organisms to specific environmental conditions. The same may be true of many key steps in vertical evolution. Think, for example, of the special conditions that may have been needed to provide a selective advantage to lungfish and their successors on the way to amphibians.

Such considerations bring to the fore the second intervention of contingency in evolution. In the eyes of many observers, the role of environmental vagaries, even more than the fortuitous nature of mutations, means that the history of life followed totally unpredictable lines, bound not to be repeated if "the tape should be allowed to play again." This point is trivial, in a way, and merely reasserts the contingency of history in general. The French philosopher Blaise Pascal already said it when he wrote that "if Cleopatra's nose had been shorter, the entire face of the Earth would have changed." Or, to put it in a more topical context, "had another of Leopold Mozart's sperm cells won the race to the egg cell produced by Anna Maria Pertl some time around May 1, 1755, some of the most beautiful music in the world would never have been written" (de Duve, 1995, p. 264).

Similarly, it has often been stated that, if a large asteroid had not fallen on the Yucatan Peninsula some 65 million years ago, dinosaurs might still be around and mammals might still lead a precarious existence in the shadow of the great reptiles; or that, if the great Rift Valley had not, about seven million years ago, split East Africa from north to south by a chasm separating the savannah from the forest, our primate ancestors might still be up in the trees, perfectly content with a 350 cm^3 brain.

In the case of individuals, the historical contingency argument is irrefutable. Cleopatra and Mozart are indubitably unique and irreplaceable. As far as major evolutionary events are concerned, however, one may wonder whether they actually owed the *fact* of their occurrence, or merely its *timing*, to chance environmental factors. It could be, for example, that mammals were bound to displace dinosaurs at some stage for a variety of reasons and that the fall of the asteroid only precipitated an issue that would have happened sooner or later. In the same vein, had not our ancestors been isolated in the savannah – assuming this factor played a key role in hominization – some other environmental circumstance might well, sooner or later, have provided an opportunity for the primate brain to express its built-in tendency to expand.

In conclusion, events that, according to the prevalent doctrine, owed their occurrence to the conjunction of two improbable chance events – the right mutation in the right place – may have to be reinterpreted in the context of the view, offered here, of organisms continually subject to a wide variety of genetic changes and waiting, so to speak, for the environment to provide conditions under which some trait will turn out to be useful. It is not unreasonable to assume, considering the manner in which Earth environments continually change, that many evolutionary steps that were favored by some special environmental conditions were bound to take place in any case, some time, somewhere.

Final Comments

The view of the history of life on Earth that emerges from the analysis offered in this book places greater emphasis on inevitability – and singularity – than that advocated by many contemporary thinkers who have reflected on the question. Starting with the universal products of cosmic chemistry, the proposed view ascribes the early development of life to strictly chemical phenomena that, by virtue of their nature, were bound to occur under the physical–chemical conditions that prevailed at the sites where they took place, leaving no room for chance.

With deterministic chemistry continuing its role up to the present, information made its entrance with replicable RNA and, later, DNA, thus introducing the joint elements of continuity and variability that form the pillars of natural selection. It is from this juncture that the proposed scenario differs most significantly from more common accounts, by the importance it accords selective optimization. This notion, which may be described as a strong version of Darwinism, adds to the classical concept of natural selection the opportunity of providing this process with a close-to-complete range of options to screen.

In its essence, the notion of selective optimization rests on the *quantitative* estimate of probabilities and opportunities. What counts is the dimension of available spaces, be they of molecular sequences or of mutations, in relation to what may be called the *coverage ability* of the evolving

238

systems. To the defenders of intelligent design, the size of those spaces exceeds coverage abilities to such an extent that the spots occupied in them by life could be reached only with the guidance of some extraneous agency of unknown nature.

According to the conventional contingency doctrine, the disparity is less prohibitive but still much too large, by far, for exhaustive exploration to be possible. Chance shots have nevertheless succeeded in many instances because a large number of spots compatible with survival existed in the spaces. From lucky hit to lucky hit, the process followed an erratic, unpredictable course, one in myriad possible pathways.

Selective optimization goes farther, postulating limited spaces and matching coverage abilities, so that unguided attainment of the best or near-best spot in the space was possible.

Let there be no mistake, I am not claiming that evolution has been entirely ruled by this mechanism, which would make its course strictly deterministic and reproducible. My models leave plenty of leeway for historical contingency, in cases of pseudo-bottlenecks, for example, and in many other cases where either selective optimization was not possible or several equivalent options were available. Such crucial phenomena as the origin of the genetic code, mimicry, or convergent evolution, do, however, indicate that the notion of selective optimization may be of wide-ranging significance. As such, it deserves at least as much attention as the more conventional explanations of evolutionary phenomena.

Implicit in the proposed view is the notion that the history of life on Earth was determined, at least in its main lines, by the environmental conditions that surrounded its birth and evolution. A corollary of this notion is that life would similarly arise and evolve wherever the same conditions as prevailed on our planet are reproduced. From which it follows that the probability of finding life and, perhaps, conscious minds elsewhere is equivalent to the probability of the necessary conditions' being realized. To the extent that this premise is accepted, the vast resources of astronomy, cosmology, and space science can be harnessed in the search for possible habitats. Unfortunately, the resources may prove insufficient and the answer unattainable.

Even with the most sophisticated techniques and instruments imaginable, unless the laws of physics can be circumvented, we shall never be able to search more than an infinitesimal fraction of the universe for signs of life and mind, or even of mere habitability. For such signs to be discovered, either the phenomena we look for must be extremely frequent or some fantastic stroke of luck has placed a second example within our reach.[1] Should the search prove negative, as it seems very likely to be, this failure would in no way provide proof of the singularity of life and mind, or even of their rarity. We shall always, when contemplating the skies, remain free to dream of "other worlds."

---

[1] I do not have in mind here Mars or other parts of the solar system, which could harbor forms of life related to Earth life. Needless to say, any trace of extraterrestrial life would be of tremendous interest, but for the finding to be relevant to the question raised, there should be proof of independent origin. Opposite chirality would be a particularly telling clue.

Bibliography

A. Akhmanova, F. Voncken, T. van Halen, A. van Hoek, B. Boxma, G. Vogels, M. Veenhuis, and H. H. P. Hackstein (1998). A hydrogenosome with a genome. *Nature*, **396**, 527–528.

A. D. Anbar and A. H. Knoll (2002). Proterozoic ocean chemistry and evolution: A bioinorganic bridge. *Science*, **297**, 1137–1142.

J. O. Andersson, A. M. Sjögren, L. A. M. Davis, T. M. Embley, and A. J. Roger (2003). Phylogenetic analyses of diplomonad genes reveal frequent lateral gene transfers affecting eukaryotes. *Curr. Biol.*, **13**, 94–104.

L. Aravind, R. L. Tatusov, Y. I. Wolf, D. R. Walker, and E. V. Koonin (1998). Evidence for massive gene exchange between archaeal and bacterial hyperthermophiles. *Trends Genet.*, **14**, 442–444.

G. L. Arnold, A. D. Anbar, J. Barling, and T. W. Lyons (2004). Molybdenum isotope evidence for widespread anoxia in mid-proterozoic oceans. *Science*, **304**, 87–90.

G. Arrhenius, B. Gedulin, and S. Mojzsis (1993). Phosphate in models for chemical evolution. In *Chemical Evolution and Origin of Life* (C. Ponnamperuma and J. Chela-Flores, eds.), 25–50. Hampton VA: A. Deepak Publishers.

M. Balter (2000). Evolution on life's fringes. *Science*, **289**, 1866–1867.

M. Baltscheffsky and H. Baltscheffsky (1992). Inorganic pyrophosphate and inorganic pyrophosphatases. In *Molecular Mechanisms in Bioenergetics* (L. Ernster, ed.), 331–348. Amsterdam: Elsevier.

A. Bar-Nun, E. Kochavi, and S. Bar-Nun (1994). Assemblies of free amino acids as possible prebiotic catalysts. *J. Mol. Evol.*, **39**, 116–122.

A. D. Baughn and M. H. Malamy (2004). The strict anaerobe *Bacteroides fragilis* grows in and benefits from nanomolar concentrations of oxygen. *Nature*, **427**, 441–444.

A. Bekker, H. D. Holland, P.-L. Wang, D. Rumble III, H. J. Stein, J. L. Hannah, L. L. Coetzee, and N. J. Beukes (2004). Dating the rise of atmospheric oxygen. *Nature*, **427**, 117–120.

M. P. Bernstein, J. P. Dworkin, S. A. Sandford, G. W. Cooper, and L. J. Allamandola (2002). Racemic amino acids from the ultraviolet photolysis of interstellar ice analogues. *Nature*, **416**, 401–403.

O. Botta and J. L. Bada (2002). Extraterrestrial organic compounds in meteorites. *Surv. Geophys.*, **23**, 411–467.

O. Botta, D. P. Glavin, G. Kminek, and J. L. Bada (2002). Relative amino acid concentrations as a signature for parent body processes of carbonaceous chondrites. *Orig. Life Evol. Biosph.*, **32**, 143–164.

Y. Boucher, M. Kamekura, and W. F. Doolittle (2004). Origins and evolution of isoprenoid lipid biosynthesis in archaea. *Mol. Microbiol.*, **52** (No. 2), 515–527.

A. Brack (2003). La chimie de l'origine de la vie. In *Les Traces du Vivant* (M. Gargaud, D. Despois, J.-P. Parisot, and J. Reisse, eds.), 61–81. Pessac: Presses Universitaires de Bordeaux.

M. D. Brasier, O. R. Green, A. P. Jephcoat, A. K. Kleppe, M. J. Van Kranendonk, J. F. Lindsay, A. Steele, and N. V. Grassineau (2002). Questioning the evidence for Earth's oldest fossils. *Nature*, **416**, 76–81.

J. J. Brocks, G. A. Logan, R. Buick, and R. E. Summons (1999). Archaean molecular fossils and the early rise of eukaryotes. *Science*, **285**, 1033–1036.

P. Brown, T. Sutikna, M. J. Morwood, R. P. Soejono, E. Wayhu Saptomo, and R. Awe Due (2004). A new small-bodied hominin from the late Pleistocene of Flores, Indonesia. *Nature*, **431**, 1055–1061.

R. Cammack (1983). Evolution and diversity in the iron–sulfur proteins. *Chem. Scr.*, **21**, 87–95.

D. E. Cane (1997). Polyketide and nonribosomal polypeptide biosynthesis. *Chem. Rev.*, **97**, 2463–2706. (includes 13 papers on the topic.)

D. E. Canfield (1998). A new model for Proterozoic ocean chemistry. *Nature*, **396**, 450–453.

L. H. Caporale (2003). Foresight in genome evolution. *Am. Sci.*, **91**, 234–241.

D. Caramelli, C. Lalueza-Fox, C. Vernes, M. Lari, A. Casoli, F. Mallegni, B. Chiarelli, I. Dupanloup, J. Bertranpetit, G. Barbujani, and G. Bertorelle (2003). Evidence for a genetic discontinuity between Neandertals and 24,000-year-old anatomically modern Europeans. *Proc. Nat. Acad. Sci. U.S.A.*, **100**, 6593–6597.

S. B. Carroll (2003). Genetics and the making of *Homo sapiens*. *Nature*, **422**, 849–857.

J. Castresana and M. Saraste (1995). Evolution of energetic metabolism: The respiration early hypothesis. *Trends Biol. Sci.*, **20**, 443–448.

T. Cavalier-Smith (1975). The origin of nuclei and eukaryotic cells. *Nature*, **256**, 463–468.

T. Cavalier-Smith (2002). The phagotrophic origin of eukaryotes and phylogenetic classification of Protozoa. *Int. J. Syst. Evol. Microbiol.*, **52**, 297–354.

T. R. Cech (1986). RNA as an enzyme. *Sci. Am.*, **255** (No. 5), 64–75.

A. Chakrabarti, R. R. Breaker, G. F. Joyce, and D. W. Deamer (1994). Production of RNA by a polymerase protein encapsulated within phospholipid vesicles. *J. Mol. Evol.*, **39**, 555–559.

S. Conway Morris (1998). *The Crucible of Creation*. Oxford: Oxford University Press.

S. Conway Morris (2003). *Life's Solution*. Cambridge: Cambridge University Press.

J. R. Cronin and S. Pizzarello (1997). Enantiomeric excesses in meteoritic amino acids. *Science*, **275**, 951–955.

C. Cunchillos and G. Lecointre (2002). Early steps of metabolism evolution inferred by cladistic analysis of amino acid catabolic pathways. *C. R. Biol.*, **325**, 119–129.

C. Cunchillos and G. Lecointre (2005). Integrating the universal metabolism into a phylogenetic analysis. *Mol. Biol. Evol.*, **22**, 1–11.

D. W. Deamer (1998). Membrane compartments in prebiotic evolution. In *The Molecular Origins of Life* (A. Brack, ed.), 189–205. Cambridge: Cambridge University Press.

C. de Duve (1963). The lysosome concept. In *Ciba Foundation Symposium on Lysosomes* (A. V. S. de Reuck and M. P. Cameron, eds.), 1–31. London: Churchill.

C. de Duve (1969). Evolution of the peroxisome. *Ann. N. Y. Acad. Sci.*, **168**, 369–381.

C. de Duve (1984). *A Guided Tour of the Living Cell*. New York: Scientific American Books.

C. de Duve (1988). The second genetic code. *Nature*, **333**, 117–118.

C. de Duve (1991). *Blueprint for a Cell*. Burlington, NC: Neil Patterson Publishers, Carolina Biological Supply Company.

C. de Duve (1993). The RNA world: Before and after? *Gene*, **135**, 29–31.

C. de Duve (1995). *Vital Dust*. New York: BasicBooks.

C. de Duve (1996). The birth of complex cells. *Sci. Am.*, **274** (No. 4), 38–45.

C. de Duve (1998). Clues from present-day biology: The thioester world. In *The Molecular Origins of Life* (A. Brack, ed.), 219–236. Cambridge: Cambridge University Press.

C. de Duve (2001). The origin of life: Energy. In *Frontiers of Life*. (D. Baltimore, R. Dulbecco, F. Jacob, and R. Levi-Montalcini, eds.), Vol. I, 153–168. San Diego, CA: Academic Press.

C. de Duve (2002). *Life Evolving*. New York: Oxford University Press.

C. de Duve (2003). A research proposal on the origin of life. *Orig. Life Evol. Biosph.*, **33**, 1–16.

C. de Duve (2005). The onset of selection. *Nature*, **433**, 581–582.

C. de Duve and R. Wattiaux (1966). Functions of lysosomes. *Annu. Rev. Physiol.*, **28**, 435–492.

F. M. Devienne, C. Barnabé, M. Couderc, and G. Ourisson (1998). Synthesis of biological compounds in quasi-interstellar conditions. *C. R. Acad. Sci. Paris*, Sér. IIc, **1**, 435–439.

F. M. Devienne, C. Barnabé, and G. Ourisson (2002). Synthesis of further biological compounds in interstellar-like conditions. *C. R. Acad. Sci. Paris, Chimie/Chemistry*, **5**, 651–653.

R. E. Dickerson and I. Geis (1969). *The Structure and Action of Proteins*. Menlo Park, CA: Benjamin/Cummings Publishing Company.

M. Di Giulio (2003). The early phases of genetic code origin: Conjectures on the evolution of coded catalysis. *Orig. Life Evol. Biosph.*, **33**, 479– 489.

W. F. Doolittle (1999). Phylogenetic classification and the universal tree. *Science*, **284**, 2124–2128.

W. F. Doolittle (2000). The nature of the universal ancestor and the evolution of the proteome. *Curr. Opin. Struct. Biol.*, **10**, 355–358.

S. D. Dyall, M. T. Brown, and P. J. Johnson (2004a). Ancient invasions: From endosymbionts to organelles. *Science*, **304**, 253–257.

S. D. Dyall, W. Yan, M. G. Delgadillo-Correa, A. Lunceford, J. A. Loo, C. F. Clarke, and P. J. Johnson (2004b). Non-mitochondrial complex I proteins in a *Trichomonas* hydrogenosomal oxidoreductase complex. *Nature*, **431**, 1103–1107.

P. Ehrenfreund, W. Irvine, L. Becker, J. Blank, J. R. Brucato, L. Colangeli, S. Derenne, D. Despois, A. Dutrey, H. Fraaije, A. Lazcano, T. Owen, and F. Robert, an International Space Science Institute ISSI Team (2002). Astrophysical and astrochemical insights into the origin of life. *Rep. Prog. Phys.*, **65**, 1427–1487.

M. Eigen and P. Schuster (1977). The hypercycle: A principle of self-organization. Part A: Emergence of the hypercycle. *Naturwissenschaften*, **64**, 541–565.

M. Eigen and R. Winkler-Oswatitsch (1981). Transfer-RNA, an early gene. *Naturwissenschaften*, **68**, 282–292.

A. Eschenmoser (1999). Chemical etiology of nucleic acid structure. *Science*, **284**, 2118–2124.

P. Forterre (1995). Thermoreduction, a hypothesis for the origin of prokaryotes. *C. R. Acad. Sci. III*, **318**, 415–422.

P. Forterre (1999). Where is the root of the universal tree of life? *BioEssays*, **21**, 871–879.

P. Forterre, C. Bouthier de la Tour, H. Philippe, and M. Duguet (2000). Reverse gyrase from hyperthermophiles: Probable transfer of a thermoadaptation trait from Archaea to Bacteria. *Trends Genet.*, **16**, 152–154.

S. W. Fox (1988). *The Emergence of Life*. New York: BasicBooks.

S. J. Freeland, T. Wu, and N. Keulmann (2003). The case for an error minimizing standard genetic code. *Orig. Life Evol. Biosph.*, **33**, 457–477.

N. Fujii (2002). D-Amino acids in living higher organisms. *Orig. Life Evol. Biosph.*, **32**, 103–127.

H. Furnes, N. R. Banerjee, K. Muehlenbachs, H. Staudigel, and M. de Wit (2004). Early life recorded in archean pillow lavas. *Science*, **304**, 578–581.

N. Galtier, N. Tourasse, and M. Gouy (1999). A nonhyperthermophilic common ancestor to extant life forms. *Science*, **283**, 220–221.

J. M. Garcia-Ruiz, S. T. Hyde, A. M. Carnerup, A. G. Christy, M. J. Van Kranendonk, and N. J. Welham (2003). Self-assembled silica–carbonate structures and detection of ancient microfossils. *Science*, **302**, 1194–1197.

B. Gedulin and G. Arrhenius (1994). Sources and geochemical evolution of RNA precursor molecules – the role of phosphate. In *Early Life on Earth*, Nobel Symposium 84 (S. Bengtson, ed.), 91–110. New York: Columbia University Press.

W. Gilbert (1986). The RNA world. *Nature*, **319**, 618.

W. Gilbert, M. Marchionni, and G. Mcknight (1986). On the antiquity of introns. *Cell*, **46**, 151–154.

N. Glansdorff (2000). About the last common ancestor, the universal life-tree, and lateral gene transfer: A reappraisal. *Mol. Microbiol.*, **38** (No. 2), 177–185.

M. Gogarten-Boekels, E. Hilario, and J. P. Gogarten (1995). The effects of heavy meteorite bombardment on the early evolution – the emergence of the three domains of life. *Orig. Life Evol. Biosph.*, **25**, 251–264.

S. Gribaldo and H. Philippe (2002). Ancient phylogenetic relationships. *Theor. Pop. Biol.*, **61**, 391–408.

R. S. Gupta, K. Aitken, M. Falah, and B. Singh (1994). Cloning of *Giardia lamblia* heat shock protein HSP70 homologs: Implications regarding origin of eukaryotic cells and endoplasmic reticulum. *Proc. Nat. Acad. Sci. U.S.A.*, **91**, 2895–2899.

J. H. P. Hackstein, A. Akhmanova, F. Voncken, A. van Hoek, T. van Alen, B. Boxma, S. Y. Moon-van der Staay, G. van der Staay, J. Leunissen, M. Huynen, J. Rosenberg, and M. Veenhuis (2001). Hydrogenosomes: Convergent adaptations of mitochondria to anaerobic environments. *Zoology*, **104**, 290–302.

M. M. Hanczyc, S. M. Fujikawa, and J. W. Szostak (2003). Experimental models of primitive cellular compartments: Encapsulation, growth, and division. *Science*, **302**, 618–622.

H. Hartman (1984). The origin of the eukaryotic cell. *Speculations Sci. Technol.*, **7** (No. 2), 77–81.

H. Hartman and A. Fedorov (2002). The origin of the eukaryotic cell: A genomic investigation. *Proc. Nat. Acad. Sci. U.S.A.*, **99**, 1420–1425.

T. Horiike, K. Hamada, S. Kanaya, and T. Shinozawa (2001). Origin of eukaryotic cell nuclei by symbiosis of Archaea in Bacteria is revealed by homology-hit analysis. *Nature Cell Biol.*, **3**, 210–214.

D. S. Horner, P. G. Foster, and T. M. Embley (2000). Iron hydrogenases and the evolution of anaerobic eukaryotes. *Mol. Biol. Evol.*, **17** (No. 11), 1695–1709.

C. Huber and G. Wächtershäuser (1998). Peptides by activation of amino acids by CO on (Ni, Fe)S surfaces: Implications for the origin of life. *Science*, **281**, 670–672.

E. Imai, H. Honda, K. Hatori, A. Brack, and K. Matsuno (1999). Elongation of oligopeptides in a simulated submarine hydrothermal system. *Science*, **283**, 831–833.

W. K. Johnston, P. J. Unrau, M. S. Lawrence, M. E. Glasner, and D. P. Bartel (2001). RNA-catalyzed RNA polymerization: Accurate and general RNA-templated primer extension. *Science*, **292**, 1319–1325.

A. Jorissen and C. Cerf (2002). Asymmetric photoreactions as the origin of biomolecular homochirality: A critical review. *Orig. Life Evol. Biosph.*, **32**, 129–142.

G. F. Joyce (2002). The antiquity of RNA-based evolution. *Nature*, **418**, 214–221.

G. F. Joyce and L. E. Orgel (1993). Prospects for the understanding of the origin of the RNA world. In *The RNA World* (R. F. Gesteland and J. F. Atkins, eds.), 1–25. Cold Spring Harbor, NY: Cold Spring Harbor Laboratory Press.

O. Kandler (1994a). Cell wall biochemistry in Archaea and its phylogenetic implications. *J. Biol. Phys.*, **20**, 165–169.

O. Kandler (1994b). The early diversification of life. In *Early Life on Earth*, Nobel Symposium 84 (S. Bengtson, ed.), 152–160. New York: Columbia University Press.

K. Kashefi and D. R. Lovley (2003). Extending the upper temperature limit for life. *Science*, **301**, 934.

M. Kates (1992). Archaebacterial lipids: Structure, biosynthesis and function. In *The Archaebacteria: Biochemistry and Biotechnology* (M. J. Danson, D. W. Hough, and G. G. Lunt, eds.), 51–72. Biochem. Soc. Symp. 58. London: Portland Press.

A. D. Keefe and S. L. Miller (1995). Are polyphosphates or phosphate esters prebiotic reagents? *J. Mol. Evol.*, **41**, 693–702.

A. D. Keefe, G. L. Newton, and S. L. Miller (1995). A possible prebiotic synthesis of pantetheine, a precursor of coenzyme A. *Nature*, **373**, 683–685.

A. H. Knoll (2003). *Life on a Young Planet*. Princeton, NJ: Princeton University Press.

V. Kolb, S. Zhang, Y. Xu, and G. Arrhenius (1997). Mineral-induced phosphorylation of glycolate ion – a metaphor in chemical evolution. *Orig. Life Evol. Biosph.*, **27**, 485–503.

E. V. Koonin (2003). Comparative genomics, minimal gene-sets and the last universal common ancestor. *Nature Rev. Microbiol.*, **1**, 127–136.

A. Kornberg, N. N. Rao, and D. Ault-Riché (1999). Inorganic polyphosphate: A molecule of many functions. *Annu. Rev. Biochem.*, **68**, 89–125.

I. S. Kulaev (1979). *The Biochemistry of Inorganic Polyphosphates*. New York: Wiley.

J. A. Lake, R. Jain, and M. C. Rivera (1999). Mix and match in the tree of life. *Science*, **283**, 2027–2028.

N. Lane (2002). *Oxygen*. Oxford: Oxford University Press.

A. Lazcano (2003). Just how pregnant is the universe? *Science*, **299**, 347–348.

L. Leman, L. Orgel, and M. Reza Ghadiri (2004). Carbonyl sulfide-mediated prebiotic formation of peptides. *Science*, **306**, 283–286.

R. Lewin (1996). *Patterns in Evolution*. New York: Scientific American Books.

M. R. Lindsay, R. I. Webb, M. Strous, M. S. Jetten, M. K. Butler, R. J. Forde, and J. A. Fuerst (2001). Cell compartmentalisation in planctomycetes: Novel types of structural organisation for the bacterial cell. *Arch. Microbiol.*, **175** (No. 6), 413–429.

P. L. Luisi (2002). Some open questions about the origin of life. In *Fundamentals of Life* (G. Palyi, C. Zucchi, and L. Caglioti, eds.), 289–301. Paris: Elsevier.

D. A. Mac Donaill (2003). Why nature chose A, C, G, and U/T: An error-coding perspective of nucleotide alphabet composition. *Orig. Life Evol. Biosph.*, **33**, 433–455.

L. Margulis (1981). *Symbiosis in Cell Evolution*. San Francisco: W. H. Freeman & Co.

L. Margulis (1996). Archaeal–eubacterial mergers in the origin of Eukarya: Phylogenetic classification of life. *Proc. Nat. Acad. Sci. U.S.A.*, **93**, 1071–1076.

L. Margulis and D. Sagan (1986). *Micro-cosmos*. New York: Summit Books.

W. Martin (1999). Mosaic bacterial chromosomes: A challenge en route to a tree of genomes. *BioEssays*, **21**, 99–104.

W. Martin and M. Müller (1998). The hydrogen hypothesis for the first eukaryote. *Nature*, **392**, 37–41.

W. Martin, C. Rotte, M. Hoffmeister, U. Theissen, G. Gelius-Dietrich, S. Ahr, and K. Henze (2003). Early cell evolution, eukaryotes, anoxia, sulfide, oxygen, fungi first (?), and a tree of genomes revisited. *Life*, **55** (No. 4–5), 193–204.

W. Martin and M. J. Russell (2003). On the origins of cells: A hypothesis for the evolutionary transitions from abiotic geochemistry to chemoautotrophic prokaryotes, and from prokaryotes to nucleated cells. *Phil. Trans. R. Soc. London B*, **358**, 59–85.

S. L. Miller (1953). A production of amino acids under possible primitive earth conditions. *Science*, **117**, 528–529.

K. Miller (1999). *Finding Darwin's God*. New York: HarperCollins.

M. Mirazon Lahr and R. Foley (2004). Human evolution writ small. *Nature*, **431**, 1043–1044.

P. A. Monnard, C. L. Apel, A. Kanavarioti, and D. W. Deamer (2004). Influence of ionic inorganic solutes on self-assembly and polymerization processes related to early forms of life: Implications for a prebiotic aqueous medium. *Astrobiology*, **2** (No. 2), 139–152.

D. Moreira and P. Lopez-Garcia (1998). Symbiosis between methanogenic archaea and δ-proteobacteria as the origin of eukaryotes: The syntrophic hypothesis. *J. Mol. Evol.*, **47**, 517–530.

H. J. Morowitz (1999). A theory of biochemical organization, metabolic pathways, and evolution. *Complexity*, **4** (No. 6), 39–53.

M. J. Morwood, R. P. Soejono, R. G. Roberts, T. Sutikna, C. S. M. Turney, K. E. Westaway, W. J. Rink, J.-x. Zhao, G. D. van den Bergh, R. Awe Due, D. R. Hobbs, M. W. Moore, M. I. Bird, and L. K. Fifield (2004). Archaeology and age of a new hominin from Flores in eastern Indonesia. *Nature*, **431**, 1087–1091.

M. Müller (1993). The hydrogenosome. *J. Gen. Microbiol.*, **139**, 2879–2889.

M. Müller and W. Martin (1999). The genome of *Rickettsia prowazekii* and some thoughts on the origin of mitochondria and hydrogenosomes. *BioEssays*, **21**, 377–381.

G. M. Munoz Caro, U. J. Meierhenrich, W. A. Schutte, B. Barbier, A. Arcones Segovia, H. Rosenbauer, W. H.-P. Thiemann, A. Brack, and J. M. Greenberg (2002). Amino acids from ultraviolet irradiation of interstellar ice analogues. *Nature*, **416**, 403–406.

K. E. Nelson, R. A. Clayton, S. R. Gill, M. L. Gwinn, R. J. Dodson, D. H. Haft, E. K. Hickey, J. D. Peterson, W. C. Nelson, K. A. Ketchum, L. McDonald, T. R. Utterback, J. A. Malek, K. D. Linher, M. M. Garrett, A. M. Stewart, M. D. Cotton, M. S. Pratt, C. A. Phillips, D. Richardson, J. Heidelberg, G. G. Sutton, R. D. Fleischmann, J. A. Eisen, O. White, S. L. Salzberg, H. O. Smith, J. C. Venter, and C. M. Fraser (1999). Evidence for lateral gene transfer between Archaea and Bacteria from genome sequence of *Thermotoga maritima*. *Nature*, **399**, 323–329.

E. Nevo (1999). *Mosaic Evolution of Subterranean Mammals*. Oxford: Oxford University Press.

E. Nogales, K. H. Downing, L. A. Amos, and J. Löwe (1998). Tubulin and FtsZ form a distinct family of GTPases. *Nature Struct. Biol.*, **5**, 451–458.

H. Ochman, J. G. Lawrence, and E. A. Groisman (2000). Lateral gene transfer and the nature of bacterial evolution. *Nature*, **405**, 299–304.

H. Ogasawara, A. Yoshida, E. Imai, H. Honda, K. Hatori, and K. Matsuno (2000). Synthesizing oligomers from monomeric nucleotides in simulated hydrothermal environments. *Orig. Life Evol. Biosph.*, **30**, 519–526.

Y. Ogata, E. Imai, H. Honda, K. Hatori, and K. Matsuno (2000). Hydrothermal circulation of sea water through hot vents and contribution of interface chemistry to prebiotic synthesis. *Orig. Life Evol. Biosph.*, **30**, 527–537.

L. E. Orgel (2003). Some consequences of the RNA world hypothesis. *Orig. Life Evol. Biosph.*, **33**, 211–218.

S. Osawa (1995). *Evolution of the Genetic Code*. Oxford University Press.

G. Ourisson and T. Nakatani (1994). The terpenoid theory of the origin of cellular life: The evolution of terpenoids to cholesterol. *Chemistry and Biology*, **1**, 11–23.

K. Ozawa, A. Nemoto, E. Imai, H. Honda, K. Hatori, and K. Matsuno (2004). Phosphorylation of nucleotide molecules in hydrothermal environments. *Orig. Life Evol. Biosph.*, **34**, 465–471.

J. D. Palmer (2003). The symbiotic birth and spread of plastids: How many times and whodunit? *J. Phycol.*, **39**, 4–11.

S. Pitsch, A. Eschenmoser, B. Gedulin, S. Hui, and G. Arrhenius (1995). Mineral induced formation of sugar phosphates. *Orig. Life Evol. Biosph.*, **25**, 294–334.

S. Pizzarello and A. L. Weber (2004). Prebiotic amino acids as asymmetric catalysts. *Science*, **303**, 1151.

A. Poole, D. Jeffares, and D. Penny (1999). Early evolution: prokaryotes, the new kids on the block. *Bioessays*, **21**, 880–889.

D. Prangishvili (2003). Evolutionary insights from studies on viruses of hyperthermophilic archaea. *Res. Microbiol.*, **154**, 289–294.

B. P. Prieur (2001). Étude de l'activité prébiotique potentielle de l'acide borique. *C. R. Acad. Sci. Paris, Chimie/Chemistry*, **4**, 1–4.

A. Ricardo, M. A. Carrigan, A. N. Olcott, and S. A. Benner (2004). Borate minerals stabilize ribose. *Science*, **303**, 196.

M. C. Rivera and J. A. Lake (2004). The ring of life provides evidence for a genome fusion origin of eukaryotes. *Nature*, **431**, 152–155.

M. Rohmer (1999). The discovery of a mevalonate-independent pathway for isoprenoid biosynthesis in bacteria, algae, and higher plants. *Nat. Prod. Rep.*, **16**, 565–574.

M. Rohmer, C. Grosdemange-Billiard, M. Seemann, and D. Tritsch (2004). Isoprenoid biosynthesis as a novel target for antibacterial and antiparasitic drugs. *Curr. Opin. Investig. Drugs*, **5** (No. 2), 154–162.

L. Sagan (1967). On the origin of mitosing cells. *J. Theor. Biol.*, **14**, 225–274.

J. G. Schmidt, P. E. Nielsen, and L. E. Orgel (1997). Enantiomeric cross-inhibition in the synthesis of oligonucleotides on a nonchiral template. *J. Am. Chem. Soc.*, **119**, 1494–1495.

J. W. Schopf (1999). *Cradle of Life*. Princeton, NJ: Princeton University Press.

A. W. Schwartz (1998). Origins of the RNA world. In *The Molecular Origins of Life* (A. Brack, ed.), 237–254. Cambridge: Cambridge University Press.

M. Shimizu (1982). Molecular basis for the genetic code. *J. Mol. Evol.*, **18**, 297–303.

A. Shimoyama and R. Ogasawara (2002). Peptides and diketopiperazines in the Yamato-791198 and Murchison carbonaceous chondrites. *Orig. Life Evol. Biosph.*, **32**, 165–179.

N. H. Sleep, K. J. Zahnle, J. F. Kasting, and H. J. Morowitz (1989). Annihilation of ecosystems by large asteroid impacts on the early Earth. *Nature*, **342**, 139–142.

M. L. Sogin, J. D. Silberman, G. Hinkle, and H. G. Morrison (1996). Problems with molecular diversity in the Eukarya. In *Evolution of Microbial Life* (D. McL. Roberts, P. Sharp, G. Alderson, and M. A. Collins, eds.), 167–184. Cambridge: Cambridge University Press.

S. Spiegelman (1967). An in vitro analysis of a replicating molecule. *Am. Sci.*, **55**, 221–264.

R. Y. Stanier (1970). Some aspects of the biology of cells and their possible evolutionary significance. *Symp. Soc. Gen. Microbiol.*, **20**, 1–38.

R. Sutak, P. Dolezal, H. L. Fiumera, I. Hrdy, A. Dancys, M. Delgadillo-Correa, P. J. Johnson, M. Müller, and J. Tachezy (2004). Mitochondrial-type assembly of Fe–S centers in the hydrogenosomes of the amitochondriate eukaryote *Trichomonas vaginalis*. *Proc. Nat. Acad. Sci. U.S.A.*, **101**, 10368–10373.

E. Szathmary (2002). The gospel of inevitability. *Nature*, **419**, 779–780.

J. W. Szostak, D. P. Bartel, and P. L. Luisi (2001). Synthesizing life. *Nature*, **409**, 387–390.

K. Tamura and P. Schimmel (2004). Chiral-selective aminoacylation of an RNA minihelix. *Science*, **305**, 1253.

J. Tovar, G. Leon-Avila, L. B. Sanchez, R. Sutak, J. Tachezy, M. van der Giezen, M. Hernandez, M. Müller, and J. M. Lucocq (2003). Mitochondrial remnant organelles of *Giardia* function in iron–sulphur protein maturation. *Nature*, **426**, 172–176.

F. van den Ent, L. A. Amos, and J. Löwe (2001). Prokaryotic origin of the actin cytoskeleton. *Nature*, **413**, 39–44.

M. A. van Zullen, A. Lepland, and G. Arrhenius (2002). Reassessing the evidence for the earliest traces of life. *Nature*, **418**, 627–630.

T. Vellai and G. Vida (1999). The origin of eukaryotes: The difference between prokaryotic and eukaryotic cells. *Proc. R. Soc. London B*, **266**, 1571–1577.

F. Voncken, B. Boxma, J. Tjaden, A. Akhmanova, M. Huynen, F. Verbeek, A. G. M. Tielens, I. Haferkamp, H. E. Neuhaus, G. Vogels, M. Veenhuis, and J. H. P. Hackstein (2002). Multiple origins of hydrogenosomes: Functional and phylogenetic evidence from the ADP/ATP carrier of the anaerobic chytrid *Neocallimastix* sp. *Mol. Microbiol.*, **44** (No. 6), 1441–1454.

C. D. von Dohlen, S. Kohler, S. T. Alsop, and W. R. McManus (2001). Mealybug beta-proteobacterial endosymbionts contain gamma-proteobacterial symbionts. *Nature*, **412**, 433–436.

G. Wächtershäuser (1998). Origin of life in an iron–sulfur world. In *The Molecular Origins of Life* (A. Brack, ed.), 206–218. Cambridge: Cambridge University Press.

G. Wächtershäuser (2003). From pre-cells to Eukarya – a tale of two lipids. *Mol. Microbiol.*, **47** (No. 1), 13–22.

C. T. Walsh (2004). Polyketide and nonribosomal peptide antibiotics: Modularity and versatility. *Science*, **303**, 1805–1810.

J. Washington (2000). The possible role of volcanic aquifers in prebiologic genesis of organic compounds and RNA. *Orig. Life Evol. Biosph.*, **30**, 53–79.

A. L. Weber (2001). The sugar model: Catalysis by amines and amino acid products. *Orig. Life Evol. Biosph.*, **31**, 71–86.

F. H. Westheimer (1987). Why nature chose phosphates. *Science*, **235**, 1173–1178.

T. D. White, B. Asfaw, D. DeGusta, H. Gilbert, G. D. Richards, G. Suwa, and F. C. Howell (2003). Pleistocene *Homo sapiens* from Middle Awash, Ethiopea. *Nature*, **423**, 742–747.

J. Whitfield (2004). Born in a watery commune. *Nature*, **427**, 674–676.

C. R. Woese (1987). Bacterial evolution. *Microbiol. Rev.*, **51**, 221–271.

C. R. Woese (1998). The universal ancestor. *Proc. Nat. Acad. Sci. U.S.A.*, **95**, 6854–6859.

C. R. Woese (2000). Interpreting the universal phylogenetic tree. *Proc. Nat. Acad. Sci. U.S.A.*, **97**, 8392–8396.

C. R. Woese (2002). On the evolution of cells. *Proc. Nat. Acad. Sci. U.S.A.*, **99**, 8742–8747.

C. R. Woese (2004). A new biology for a new century. *Microbiol. Mol. Biol. Rev.*, **68**, 173–186.

C. R. Woese and G. E. Fox (1977). Phylogenetic structure of the prokaryotic domain. *Proc. Nat. Acad. Sci. U.S.A.*, **74**, 5088–5090.

J. T.-F. Wong (1975). A co-evolution theory of the genetic code. *Proc. Nat. Acad. Sci. U. S. A.*, **72**, 1909–1912.

J. T.-F. Wong (1991). Origin of genetically encoded protein synthesis: A model based on selection for RNA peptidation. *Orig. Life Evol. Biosph.*, **21**, 165–176.

J. T.-F. Wong and H. Xue (2002). Self-perfecting evolution of heteropolymer building blocks and sequences as the basis of life. In *Fundamentals of Life* (G. Palyi, C. Zucchi, and L. Caglioti, eds.), 473–494. Paris: Elsevier.

Y. Xu and N. Glansdorff (2002). Was our ancestor a hyperthermophilic procaryote? *Comp. Biochem. Physiol. Part A*, **133**, 677–688.

Y. Yamagata, H. Watanabe, M. Saitoh, and T. Namba (1991). Volcanic production of polyphosphates and its relevance to prebiotic evolution. *Nature*, **352**, 516–519.

M. Yarus (2000). RNA–ligand chemistry: A testable source for the genetic code. *RNA*, **6**, 475–484.

S. Yokoyama, A. Koyama, A. Nemoto, H. Honda, E. Imai, K. Hatori, and K. Matsuno (2003). Amplification of diverse catalytic properties of evolving molecules in a simulated hydrothermal environment. *Orig. Life Evol. Biosph.*, **33**, 589–595.

W. Zillig (1991). Comparative biochemistry of Archaea and Bacteria. *Curr. Opin. Genet. Dev.*, **1**, 544–551.

Index

Acidity. *See also* pH; Protons: described, 146–147; and selection, 166
Actin. *See* Cytoskeleton
Active transport. *See* Pumps in membrane transport
Acyl transfer, 56, 57, 61, 91
Adenine. *See* Bases, purine
Adenosine, selection of, 38–40
Adenosine monophosphate (AMP): in ATP hydrolysis, 31–34, 38; construction of, 152, 156; described, 25; functions of, 216; in group transfers, 62–64, 126; in protein synthesis, 90, 95; in RNA synthesis, 67–68; structure of, 26
Adenosine triphosphate. *See* ATP (Adenosine triphosphate)
Aerobic bacteria, electron transport chains in, 135–137, 143
Aerobic defined, 44
Aerobic environments, development of, 208
Aerobic hydrogenosomes, function of, 205
Aerobic organisms: ATP assembly in, 61, 206; molecular oxygen production in, 141; and oxygen toxicity, 200
α-ketoglutarate: in ATP assembly, 58, 61; oxidative decarboxylation of, 58, 60, 61
Alcohols, long-chain in cellular membranes, 126–128. *See also* Ethanol
Amino acids: D-Amino acids, in cell walls, 12, in meteorites, 13, in murein, 176,

selection of, 96; L-Amino acids, in cell walls, 12, in meteorites, 13, in murein, 176, selection of, 96; proteinogenic, 13, 96, 103, 109; R groups of, 89, 102, 122; RNA and, 96, 103; selection of, 95, 96; sequence and function, 23, 89
Amino acyl-tRNA synthetases, 92, 103, 111
AMP. *See* Adenosine monophosphate (AMP)
Amphibians, evolution of, 218
Amphiphilic property defined, 118, 119
Anaerobic: defined, 44; fermentation, 58; hydrogen generation, 207; respiration, development of, 148
Anaerobic bacteria: ATP production in, 206; electron transport chains in, 135–137, 143; and mitochondrial development, 204
Anaerobic environments: Archaea in, 175; development of, 208; evolution of, 199–202
Ancestors of humans, 222–223
Animals: cell structure of, 171; drug resistance in, 234; electron transfer in, 44; evolution of, 186, 214, 218–219, 227; genetic code in, 100, 161; in hydrothermal vents, 156; light production in, 28; phylogenetic tree, 175; visual machinery in, 146
Anticodons, 71, 92–95, 97, 101–104

Printed in the United States
by Baker & Taylor Publisher Services